Undergraduate Texts in Mathematics

For other titles Published in this series, go to
www.springer.com/series/666

Matthias Beck • Ross Geoghegan

The Art of Proof

Basic Training for Deeper Mathematics

Springer

Matthias Beck
Department of Mathematics
San Francisco State University
San Francisco, CA 94132
USA
beck@math.sfsu.edu

Ross Geoghegan
Department of Mathematical Sciences
Binghamton University
State University of New York
Binghamton, NY 13902
USA
ross@math.binghamton.edu

ISSN 0172-6056
ISBN 978-1-4419-7022-0 e-ISBN 978-1-4419-7023-7
DOI 10.1007/978-1-4419-7023-7
Springer New York Dordrecht Heidelberg London

Library of Congress Control Number: 2010934105

Mathematics Subject Classification (2010): 00A05, 00A35

Printed on acid-free paper

Springer is part of Springer Science+Business Media (www.springer.com)

Great teachers introduced us to the arts of mathematics and writing:

To Harald Kohl and Hartmut Stapf

To the memory of
Fr. Harry Lawlor, SJ and Fr. Joseph Veale, SJ

Preface

We have written this book with several kinds of readers in mind:

(a) Undergraduates who have taken courses such as calculus and linear algebra, but who are not yet prepared for upper-level mathematics courses. We cover mathematical topics that these students should know. The book also provides a bridge to the upper-level courses, since we discuss formalities and conventions in detail, including the axiomatic method and how to deal with proofs.

(b) Mathematics teachers and teachers-in-training. We present here some of the foundations of mathematics that anyone teaching mathematics beyond the most elementary levels should know.

(c) High-school students with an unusually strong interest in mathematics. Such students should find this book interesting and (we hope) unconventional.

(d) Scientists and social scientists who have found that the mathematics they studied as undergraduates is not sufficient for their present needs. Typically, the problem here is not the absence of training in a particular technique, but rather a general feeling of insecurity about what is correct or incorrect in mathematics, a sense of material only partly understood. Scientists must be confident that they are using mathematics correctly: fallacious mathematics almost guarantees bad science.

In so far as possible we try to "work in" the formal methods indirectly, as we take the reader through some interesting mathematics. Our subject is number systems: the integers and the natural numbers (that's the discrete Part I), the real numbers and the rational numbers (the continuous Part II). In this there is emphasis on induction, recursion, and convergence. We also introduce cardinal number, a topic that links the discrete to the continuous.

We teach method: how to organize a proof correctly, how to avoid fallacies, how to use quantifiers, how to negate a sentence correctly, the axiomatic method, etc. We assert that computer scientists, physicists, mathematics teachers, mathematically inclined economists, and biologists need to understand these things. Perhaps you too if you have read this far.

We sometimes hear students speak of "theoretical math," usually in a negative tone, to describe mathematics that involves theorems and proofs rather than computations and applications. The trouble with this is that, sooner or later, mathematics becomes sufficiently subtle that fundamentals have to be understood. "[W]e share the view that applied mathematics may not exist—only applied mathematicians" (R. C. Buck, Preface to *Advanced Calculus*).

We sometimes hear students say, "I like math but I don't like proofs." They have not yet realized that a proof is nothing more than an explanation of why a carefully worded statement is true. The explanation too should be carefully worded: what is said should be what is meant and what is meant should be what is said.

But who needs that level of precision? The answer is that almost all users of mathematics, except perhaps users at purely computational levels, need to understand what they are doing, if only to have confidence that they are not making mistakes. Here are some examples.

- Every mathematically trained person should understand induction arguments and recursive definitions. It is hard to imagine how one could write a nontrivial computer program without this basic understanding. Indeed, a software engineer of our acquaintance tells us that his (small) company's software has 1.5 million lines of code, which must be easy to manage; therefore recursive algorithms are forbidden unless very clearly marked as such, and most of his programmers do not understand recursion deeply enough that their recursive programs can be trusted to be error-free: so they just insert a recursion package taken from a software library.

  You had better be clear about what axioms you are assuming for the integers and natural numbers, something discussed in detail in this book.

  Here is an algorithm problem: You have known since childhood how to add a column of many-digit numbers. Certainly, you normally do this in base 10. Can you write down, as a formally correct recursion, the algorithm you learned as a child for addition of a column of base-10 whole numbers? Your algorithm should be such that, in principle, the input can be any finite list of whole numbers, and the output should print out the digits of their sum. And (now the challenging part) once you have done this, can you prove that your algorithm always gives the correct answer? Do you even know what such a question means?

- Here is a simple probability question: A deck of n different cards is shuffled and laid on the table by your left hand, face down. An identical deck of cards, independently shuffled, is laid at your right hand, also face down. You start turning up cards at the same rate with both hands, first the top card from both decks, then the next-to-top cards from both decks, and so on. What is the probability that you will simultaneously turn up identical cards from the two decks? The answer should depend on n. As n gets very large what happens to this probability? Does it converge to 0? Or to 1? Or to some number in between? And if so, what is that number? And what exactly is meant by saying that this number is the limit of the probabilities as n gets larger and larger? And how fast (in terms of n) does the nth probability approach this limiting number?

We learned of this neat example from Persi Diaconis.

Our point is not that you should solve this little problem—though it is fun to do and not hard—but that you should be able to say with confidence that in principle you understand all the questions raised. If you cannot say that, you may need (something like) this book.

- We have heard students say: "Only the integers, and perhaps the rational numbers, have any relevance in the world; the irrational real numbers are artificial academic constructs. Why, you can't even write down their decimal expansions." This is only true in the narrowest of senses. Against the notion that irrational numbers do not appear in real life we offer:

  - The diagonal of a square of side one foot has length $\sqrt{2}$ feet.
  - The ratio of the circumference of a circle to its diameter is π.
  - The answer to our limit problem about the two decks of cards is $1 - \frac{1}{e}$. One encounters e also in mortgage calculations and exponential growth or decay.

  Besides this, irrational numbers often have to be approximated by rationals up to some specified error. How is one to do this without an understanding of the issues involved in approximation: algorithms and computation of error?

- There is an old joke among physicists that "All series converge uniformly and absolutely everywhere." Often, a physics instructor will disregard questions of convergence. For example, all terms in a power series beyond the first or second order will be discarded on the assumption that they are too small to influence the answer significantly. This works in classical situations in which the series under discussion has been known for many years to give physically plausible answers in line with often-measured experimental data (and the instructor either knows what the convergence situation is, or knows that others have checked it carefully). But *your* knowledge should not be so weak that you are not sure whether your series "converges" or "converges absolutely" or "converges uniformly," and what the difference between these is.

If any of these examples seem intriguing to you, this book was written for you.

Acknowledgments

This book is a development of class notes Ross Geoghegan has been using and altering for the past twenty-five years. He experienced a version of the Moore Method of *learning through discovery* as a graduate student, and its influence has stuck.

We thank Laura Anderson, Federico Ardila, Matthew Brin, Fernando Guzmán, Eric Hayashi, Paul Loya, Marcin Mazur, David Meredith, Gene Murrow, Dennis Pixton, Wendell Ressler, Dmytro Savchuk, James T. Smith, Marco Varisco, Diane Vavrichek, and Thomas Zaslavsky who tried out variants on these notes and generously gave us much useful feedback and suggestions for further material. Louis McAuley and the late Craig Squier gave early suggestions that influenced how the initial class notes evolved. We are grateful to the students of the classes in which we and our colleagues could try out this material. We thank Kelley Walker, who wrote many of the solutions for the "instructor's edition" of this book. Matthias Beck would like to express his gratitude to Tendai Chitewere for her love and support.

We acknowledge support from the National Science Foundation. We are grateful to Sheldon Axler for his suggestions, especially for pointing out Tom Sgouros and Stefan Ulrich's `mparhack` LaTeX package which makes marginal notes easy. We thank our editors at Springer: Hans Koelsch, Ann Kostant, and Katie Leach. Special thanks go to David Kramer for his impeccable copyediting of our manuscript and to Rajiv Monsurate for sharing his LaTeX expertise. We thank ScienceCartoonsPlus.com, United Feature Syndicate, and Universal Uclick for granting us the use of the cartoons appearing in this book.

Naturally, we take sole responsibility for all mistakes and would like to hear about any corrections or suggestions. We will collect corrections, updates, etc., at the Internet website `math.sfsu.edu/beck/aop.html`.

San Francisco
Binghamton
May 2010

Matthias Beck
Ross Geoghegan

Contents

Notes for the Student

You have been studying important and useful mathematics since the age of three. Most likely, the body of mathematics you know can be described as Sesame-Street-through-calculus. This is all good and serious mathematics—from the beautiful algorithm for addition, which we all learned in elementary school, through high-school algebra and geometry, and on to calculus.

Now you have reached the stage where the details of what you already know have to be refined. You need to understand them from a more advanced point of view. We want to show you how this is done. We will take apart what you thought you knew (adding some new topics when it seems natural to do so) and reassemble it in a manner so clear that you can proceed with confidence to deeper mathematics—algebra, analysis, combinatorics, geometry, number theory, statistics, topology, etc.

Actually, we will not be looking at everything you know—that would take too long. We concentrate here on numbers: integers, fractions, real numbers, decimals, complex numbers, and cardinal numbers. We wish we had time to do the same kind of detailed examination of high-school geometry, but that would be another book, and, as mathematical training, it would only teach the same things over again. To put that last point more positively: once you understand what we are teaching in this book—in this course—you will be able to apply these methods and ideas to other parts of mathematics in future courses.

The topics covered here form part of the standard "canon" that everyone trained in mathematics is assumed to know. Books on the history of mathematics, for example, *Mathematics and Its History* (by J. Stillwell, Springer, 2004) and *Math Through the Ages* (by W. P. Berlinghoff and F. Q. Gouvea, Oxton House, 2002), discuss who first discovered or introduced these topics. Some go back hundreds of years; others were developed gradually, and reached their presently accepted form in the early twentieth century. We should say clearly that no mathematics in this book originates with us.

On first sight you may find this book unusual, maybe even alarming. Here is one comment we received from a student who used a test version:

> The overall feel of the book is that it is very "bare bones"; there isn't much in the way of any additional explanations of any of the concepts. While this is nice in the sense that the definitions and axioms are spelled right out without anything getting in the way, if a student doesn't initially understand the concepts underlying the sentence, then they're screwed. As it stands, the book seems to serve as a supplement to a lecture, and not entirely as a stand-alone learning tool.

This student has a point, though we added more explanations in response to comments like this. We intend this book to be supplemented by discussion in an instructor's class. If you think about what is involved in writing any book of instruction you will realize that the authors had better be clear about the intended readership and the way they want the book to be used. While we do believe that *some* students can use this book for self-study, our experience in using this material—experience stretching over twenty-five years—tells us that this will not work for everyone. So please regard your instructor as Part 3 of this book (which comes in two parts), as the source for providing the insights we did not—indeed, could not—write down.

We are active research mathematicians, and we believe, for ourselves as well as for our students, that learning mathematics through oral discussion is usually easier than learning mathematics through reading, even though reading and writing are necessary in order to get the details right. So we have written a kind of manual or guide for a semester-long discussion—inside and outside class.

Please read the *Notes for Instructors* on the following pages. There's much there that's useful for you too. And good luck. Mathematics is beautiful, satisfying, fun, deep, and rich.

Notes for Instructors

Logic moves in one direction, the direction of clarity, coherence and structure. Ambiguity moves in the other direction, that of fluidity, openness, and release. Mathematics moves back and forth between these two poles. [...] It is the interaction between these different aspects that gives mathematics its power.
William Byers (*How Mathematicians Think*, Princeton University Press, 2007)

This book is intended primarily for students who have studied calculus or linear algebra and who now wish to take courses that involve theorems and proofs in an essential way. The book is also for students who have less background but have strong mathematical interests.

We have written the text for a one-semester or two-quarter course; typically such a course has a title like "Gateway to Mathematics" or "Introduction to proofs" or "Introduction to Higher Mathematics." Our book is shorter than most texts designed for such courses. Our belief, based on many years of teaching this type of course, is that the roles of the instructor and of the textbook are less important than the degree to which the student is invited/requested/required to do the hard work.

Here is what we are trying to achieve:

1. To show the student some important and interesting mathematics.
2. To show the student how to read and understand statements and proofs of theorems.
3. To help the student discover proofs of stated theorems, and to write down the newly discovered proofs correctly, and in a professional way.
4. To foster in the student something as close as feasible to the experience of doing research in mathematics. Thus we want the student to actually discover theorems and write down correct and professional proofs of those discoveries. This is different from being able to write down proofs of theorems that have been pre-certified as true by us (in the text) or by the instructor (in class).

Once the last of these has been achieved, the student is a mathematician. We have no magic technique for getting the student to that point quickly, but this book might serve as a start.

Many books intended for a gateway course are too abstract for our taste. They focus on the different types of proofs and on developing techniques for knowing when to use each method. We prefer to start with useful mathematics on day one, and to let the various methods of proof, definition, etc., present themselves naturally as they are needed in context.

Here is a quick indication of our general philosophy:

On Choice of Material

We do not start with customary dry chapters on "Logic" and "Set Theory." Rather we take the view that the student is intelligent, has considerable prior experience with mathematics, and knows, from common sense, the difference between a logical deduction and a piece of nonsense (though some training in this may be helpful!). To defuse fear from the start, we tell the student, "A theorem is simply a sentence expressing something true; a proof is just an explanation of why it is true." Of course, that opens up many other issues of method, which we gradually address as the course goes on.

We say to the student something like the following: "You have been studying important and useful mathematics since the age of three; the body of mathematics you know is Sesame-Street-through-calculus. Now it's time to revisit (some of) that good mathematics and to get it properly organized. The very first time most of you heard a theorem proved was when you asked some adult, Is there a biggest number? (What answer were you given? What would you answer now if a four-year-old asked you that question?) Later on, you were taught to represent numbers in base 10, and to add and multiply them. Did you realize how much is buried behind that (number systems, axioms, algorithms, ...)? We will take apart what you thought you knew, and we will reassemble it in a manner so clear that you can proceed with confidence to deeper mathematics."

The Parts of the Book

The material covered in this book consists of two parts of equal size, namely a discrete part (integers, induction, modular arithmetic, finite sets, etc.) and a continuous part (real numbers, limits, decimals, infinite cardinal numbers, etc.) We recommend that both parts be given equal time. Thus the instructor should resist the temptation to let class discussion of Part 1 slide on into the eighth week of a semester. Some discipline concerning homework deadlines is needed at that point too, so that students will give enough time and attention to the second half. (The instructor who ignores this advice will probably come under criticism from colleagues: this course is often a prerequisite for real analysis.) Still, an instructor has much freedom on how to go through the material. For planning purposes, we include below a diagram showing the section dependencies.

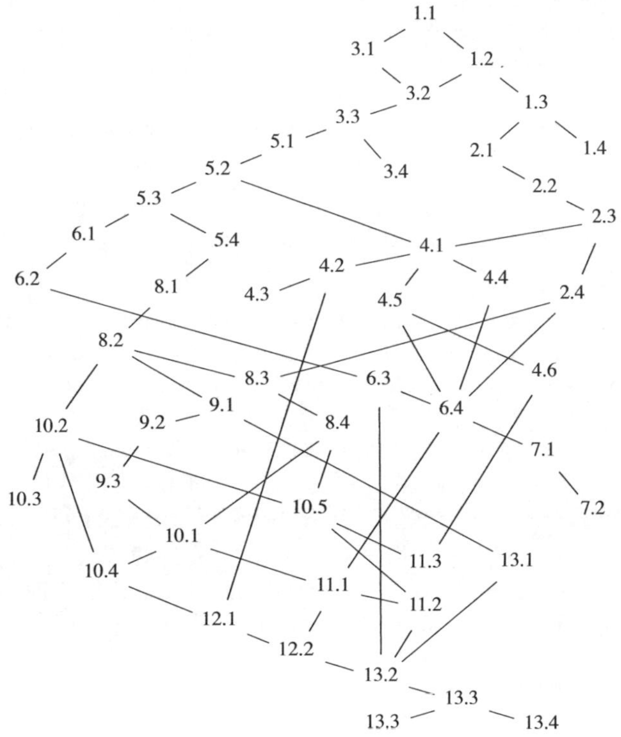

Fig. 0.1 The partially ordered set of section dependencies.

To add flexibility and material for later reading, we end the book with a collection of further topics, for example Cayley graphs of groups and public-key encryption. These additional chapters are independent of each other and can be inserted in the course as desired. They should also be suitable for student presentations in class.

Problems

There are three kinds of exercises for students in this book:

1. The main body of the text consists of *propositions* (called *theorems* when they are particularly important), in which the mathematics is developed. In principle, these propositions are meant to be proved by the students; however, proving *all* of them is likely to be overwhelming, so the instructor must exercise judgment. Besides this, some of the propositions are proved in the text to give the student a feel for how to develop a proof of a certain statement, and also to introduce different proof methods. Of the remaining propositions, we tend to prove roughly half in class on the blackboard and give the other half as homework problems.

   Upon request (see `www.springer.com/instructors`), the instructor can obtain a free copy of this book (in PDF format) in which most proofs are worked out in detail.

2. There are also exercises called *projects*. These are more exploratory, sometimes open-ended, problems for the students to work on. They vary greatly in difficulty—some are more elementary than the propositions, some concern unsolved conjectures, and some are writing projects intended to foster exploration by the students. We would encourage students to do these in groups. Some could be the basis for an outside-class pizza party, one project per party. The further topics at the end of the book also lend themselves to group projects.
3. We start every chapter with an introductory project labeled *Before You Get Started*. These are meant to be more writing intensive than the *projects* in the main text. They typically invite the students to reflect on what they already know from previous classes as a lead-in to the chapter. These introductory projects encourage the student to be creative by thinking about a topic before formally studying it.

On Grading Homework—The Red-Line Method

It is essential that the student regularly hand in written work and get timely feedback. One method of grading that we have found successful lessens the time-burden on the instructor and puts the responsibility on the shoulders of the student. It works like this:

Certain theorems in the book are assigned by the instructor: proofs are to be handed in. The instructor reads a proof until a (real) mistake is found—this might be a sentence that is false or a sentence that has no meaning. The instructor draws a red line under that sentence and returns the proof to the student at the next meeting. No words are written on the paper by the instructor: it is the student's job to figure out why the red line was put there. Pasting as necessary so as not to have to rewrite the correct part—the part above the red line—the student then hands in a new version, and the process of redlining is repeated until the proof is right.

The instructor will decide on the details of this method: how many rewrites to allow, and whether to give the same credit for a successful proof on the sixth attempt as on the first. Another issue that arises is how to handle students' questions about red lines in office hours. Some instructors will want to explain to the students why the red line was drawn. Another approach, which we have found successful, is to have the student read the proof aloud, sentence by sentence. Almost always, when the student reaches the redlined sentence it becomes clear what the issue is.

In all this we are not looking for perfection of expression—that will hopefully come with time. We start with the attitude that a proof is just an explanation of why something is true, and the student should come to understand that a confused explanation is no more acceptable in mathematics than in ordinary life. But the red line should be reserved for real mistakes of thought. To put this another way, the student needs to believe that writing correct mathematics is not an impossible task. We should be teaching rigor, but not *rigor mortis*.

We sometimes say in class that we will read the proof as if it were a computer program: if the program does not run, there must be some first line where the trouble occurs. That is where the red line is.

Part I: The Discrete

Chapter 1
Integers

CALVIN AND HOBBES © Bill Watterson. Dist. by UNIVERSAL UCLICK. Reprinted with permission.

Before You Get Started. You have used numbers like $1,2,3,34,101$, etc., ever since you learned how to count. And a little later you also met numbers like $0,-11,-40$. You know that these numbers—they are called *integers*—come equipped with two operations, "plus" and "times." You know some of the properties of these operations: for example, $3+5=5+3$ and, more generally, $m+n=n+m$. Another example: $3\cdot 5\cdot 7$ is the same whatever the order of multiplying, or, more abstractly, $(k\cdot m)\cdot n = k\cdot(m\cdot n)$. List seven similar examples of properties of the integers, things you know are correct. Are there some of these that can be derived from others? If so, does that make some features on your list more fundamental than others? In this chapter we organize information about the integers, making clear what follows from what.

M. Beck and R. Geoghegan, *The Art of Proof: Basic Training for Deeper Mathematics*, Undergraduate Texts in Mathematics, DOI 10.1007/978-1-4419-7023-7_1,

If you open a mathematics book in the library, you usually will not see a list of axioms on the first page, but they are present implicitly: the author is assuming knowledge of more basic mathematics that rests on axioms known to the reader.

We begin by writing down a list of properties of the integers that your previous experience will tell you ought to be considered to be true, things you always believed anyway. We call these properties *axioms*. Axioms are statements that form the starting point of a mathematical discussion; items that are assumed (by an agreement between author and reader) without question or deeper analysis. Once the axioms are settled, we then explore how much can be logically deduced from them. A mathematical theory is rich if a great deal can be deduced from a few primitive (and intuitively acceptable) axioms.

In short, we have to start somewhere. The axioms in one course may in fact be theorems in a deeper course whose axioms are more primitive. The list of axioms is simply a clearly stated starting point.

1.1 Axioms

You might ask, how is a set defined? We will use the word intuitively: a set is a collection of "things" or elements or members. We will say more about this in Chapter 5.

We assume there is a set, denoted by $\mathbf{Z}$, whose members are called **integers**. This set $\mathbf{Z}$ is equipped with binary operations called **addition**, $+$, and **multiplication**, $\cdot$, satisfying the following five axioms, as well as Axioms 2.1 and 2.15 to be introduced in Chapter 2. (A **binary operation** on a set S is a procedure that takes two elements of S as input and gives another element of S as output.)

Axiom 1.1. *If m, n, and p are integers, then*

(i) $m+n=n+m$. *(commutativity of addition)*

(ii) $(m+n)+p=m+(n+p)$. *(associativity of addition)*

(iii) $m\cdot(n+p)=m\cdot n+m\cdot p$. *(distributivity)*

(iv) $m\cdot n=n\cdot m$. *(commutativity of multiplication)*

(v) $(m\cdot n)\cdot p=m\cdot(n\cdot p)$. *(associativity of multiplication)*

The right-hand side of (iii) should read $(m\cdot n)+(m\cdot p)$. It is a useful convention to always multiply before adding, whenever an expression contains both $+$ and $\cdot$ (unless this order is overridden by parentheses).

Axiom 1.2. *There exists an integer* 0 *such that whenever $m\in\mathbf{Z}$, $m+0=m$.* *(identity element for addition)*

Axiom 1.3. *There exists an integer* 1 *such that $1\neq 0$ and whenever $m\in\mathbf{Z}$, $m\cdot 1=m$.* *(identity element for multiplication)*

Axiom 1.4. *For each $m\in\mathbf{Z}$, there exists an integer, denoted by $-m$, such that $m+(-m)=0$.* *(additive inverse)*

Axiom 1.5. *Let m, n, and p be integers. If $m\cdot n=m\cdot p$ and $m\neq 0$, then $n=p$.* *(cancellation)*

The symbols $\in$ and $=$. The symbol $\in$ means **is an element of**—for example, $0 \in \mathbf{Z}$ means "0 is an element of the set $\mathbf{Z}$." The symbol "$=$" means **equals**. To say $m = n$ means that m and n are the same number. We note some properties of the symbol "$=$":

In other textbooks, (i)–(iv) might form another axiom, alongside axioms for sets. In order to get to interesting mathematics early on, we chose not to include axioms on set theory and logic but count on your intuition for what a "set" should be and what it means for two members of a set to be equal.

(i) $m = m$. *(reflexivity)*

(ii) If $m = n$ then $n = m$. *(symmetry)*

(iii) If $m = n$ and $n = p$ then $m = p$. *(transitivity)*

(iv) If $m = n$, then n can be substituted for m in any statement without changing the meaning of that statement. *(replacement)*

An example of (iv): If we know that $m = n$ then we can conclude that $m + p = n + p$.

The symbol "$\neq$" means **is not equal to**. To say $m \neq n$ means m and n are different numbers. Note that "$\neq$" satisfies symmetry, but not transitivity and reflexivity.

Similarly, the symbol $\notin$ means **is not an element of**.

1.2 First Consequences

At this point, the only facts we consider known about the integers are Axioms 1.1–1.5. In the language of mathematics, the axioms are **true** or are **facts**. Every time we prove that some statement follows logically from the axioms we are proving that it too is true, just as true as the axioms, and from then on we may add it to our list of facts. Once we have established that the statement is a fact (i.e., is true) we may use it in later logical arguments: it is as good as an axiom because it follows from the axioms.

What is truth? That is for the philosophers to discuss. Mathematicans try to avoid such matters by the axiomatic method*: in mathematics a statement is considered true if it follows logically from the agreed axioms.*

From now on, we will use the common notation mn to denote $m \cdot n$. We start with some propositions that show that our axioms still hold when we change the orders of some terms:

Proposition 1.6. *If m, n, and p are integers, then $(m+n)p = mp+np$.*

Here is a proof of Proposition 1.6. Let $m, n, p \in \mathbf{Z}$. The left-hand side $(m+n)p$ of what we are trying to prove equals $p(m+n)$ by Axiom 1.1(iv). Now we may use Axiom 1.1(iii) to deduce that $p(m+n) = pm + pn$. Finally, we use Axiom 1.1(iv) again: $pm = mp$ and $pn = np$. In summary we have proved:

$$(m+n)p \overset{\text{Ax.1.1(iv)}}{=} p(m+n) \overset{\text{Ax.1.1(iii)}}{=} pm+pn \overset{\text{Ax.1.1(iv)}}{=} mp+np,$$

that is, $(m+n)p = mp+np$. □

We use □ to mark the end of a proof.

All we have used are some statements that we know to be true (Axioms 1.1(iii) and (iv)), and we mixed those together in a way that provided us with the statement in Proposition 1.6.

Have a look at the setup of this proof: We assumed we were given integers m, n, p, and using the axioms, we reached the statement $(m+n)p = mp+np$. We think of the last line of the proof as the *goal* of our work, and it is usually a good idea to write down this goal *before* showing how to get from what is given to what is to be proved.

What does it mean to "prove if $\heartsuit$ then $\clubsuit$"? The statement "if $\heartsuit$ then $\clubsuit$" might be true but not obvious; the question is how you get from $\heartsuit$ to $\clubsuit$. That journey is called a *proof* of the statement "if $\heartsuit$ then $\clubsuit$." It means: You begin by assuming $\heartsuit$. You notice that, since $\heartsuit$ is true, $\heartsuit'$ must also be true. This, in turn, makes it clear to you that $\heartsuit''$ must be true. And so on ..., where in the last step you see that $\clubsuit$ must be true.

You can prove the next propositions in a similar way; try it.

Proposition 1.7. *If m is an integer, then $0+m = m$ and $1 \cdot m = m$.*

Proposition 1.8. *If m is an integer, then $(-m)+m = 0$.*

Proposition 1.9. *Let m, n, and p be integers. If $m+n = m+p$, then $n = p$.*

Proof. Let m, n, and p be integers and $m+n = m+p$. We can add $-m$ to each side:

$$(-m)+(m+n) = (-m)+(m+p).$$

It remains to use Axiom 1.1(ii), Proposition 1.8, and Proposition 1.7 on both sides of this equation:

$$\begin{aligned} ((-m)+m)+n &= ((-m)+m)+p \\ 0+n &= 0+p \\ n &= p. \end{aligned}$$

□

What to say and what to omit. Take a careful look at the proofs we have given so far, those of Propositions 1.6 and 1.9. In the first, we stated every use of the axioms explicitly. In the second, we indicated which axioms and propositions we were using but we left it to you, the reader, to see exactly how. This suggests the question, "How much do I need to say in my proofs?" It is not an easy question to answer because it depends on two variables: the level of mathematical understanding of (a) the writer and (b) the reader. As a practical matter, the reader of your proofs in this course will be the instructor, who may be assumed to have a deep grasp of mathematics. You, the writer, are learning, so at the start, i.e., for proofs in this Chapter 1, you are advised to say everything; in other words, give details as in our proof of Proposition 1.6.

As you do this you will see that it is time-consuming and boring. You will think, "So much of what I'm doing follows obviously from the axioms and propositions, so I shouldn't have to spell it all out." You are right in the long run, but (as one of our teachers said to us once) "in mathematics, you have to earn the right to be vague." So we advise you to practice with the details until it is clear to you and your instructor what can be omitted. But this rule remains: you must say enough that both you (the writer) and your reader can see that your argument is correct and properly thought through. That part will never change.

Proposition 1.10. *Let $m, x_1, x_2 \in \mathbf{Z}$. If m, x_1, x_2 satisfy the equations $m + x_1 = 0$ and $m + x_2 = 0$, then $x_1 = x_2$.*

This means that, given $m \in \mathbf{Z}$, the integer $-m$ mentioned in Axiom 1.4 is the unique solution of the equation $m + x = 0$.

Proposition 1.11. *If m, n, p, and q are integers, then*

(i) $(m+n)(p+q) = (mp+np)+(mq+nq)$.

(ii) $m+(n+(p+q)) = (m+n)+(p+q) = ((m+n)+p)+q$.

(iii) $m+(n+p) = (p+m)+n$.

(iv) $m(np) = p(mn)$.

(v) $m(n+(p+q)) = (mn+mp)+mq$.

(vi) $(m(n+p))q = (mn)q+m(pq)$.

Why do we care about proofs? To prove a statement means convincing yourself or your audience beyond doubt that the statement is true. A proven statement is a new fact. Mathematics is like a building under construction: every new proven fact is a new brick. You do not want any defective bricks.

Here are some propositions that refine our knowledge about 0 and 1:

Proposition 1.12. *Let $x \in \mathbf{Z}$. If x has the property that for each integer m, $m + x = m$, then $x = 0$.*

Proposition 1.12 says that the integer 0 mentioned in Axiom 1.2 is the unique solution of the equation $m + x = m$.

Proposition 1.13. *Let $x \in \mathbf{Z}$. If x has the property that there exists an integer m such that $m + x = m$, then $x = 0$.*

Proposition 1.14. *For all $m \in \mathbf{Z}$, $m \cdot 0 = 0 = 0 \cdot m$.*

The propositions in this chapter are meant to be proved in the order they are presented here.

When m and n are integers, we say m **is divisible by** n (or alternatively, n **divides** m) if there exists $j \in \mathbf{Z}$ such that $m = jn$. We use the notation $n \mid m$.

Do not confuse this with the notations $\frac{n}{m}$ and n/m for fractions.

Example 1.15. You have thought about divisibility in elementary school (before you could divide two numbers). Most likely the first instance was given by **even** integers, which are defined to be those integers that are divisible by 2.

Here we define $2 = 1 + 1$. We will say more about this in the next chapter.

Proposition 1.16. *If m and n are even integers, then so are $m+n$ and mn.*

Proposition 1.17.

(i) 0 *is divisible by every integer.*

(ii) *If m is an integer not equal to* 0*, then m is not divisible by* 0.

Thus the integer 1 mentioned in Axiom 1.3 is the unique solution of the equation $mx = m$.

Proposition 1.18. *Let $x \in \mathbf{Z}$. If x has the property that for all $m \in \mathbf{Z}$, $mx = m$, then $x = 1$.*

Proposition 1.19. *Let $x \in \mathbf{Z}$. If x has the property that for some nonzero $m \in \mathbf{Z}$, $mx = m$, then $x = 1$.*

This is another if–then statement: *if* statement ♡ is true *then* statement ♣ is true as well. Statement ♡ here is "x has the property that for some nonzero $m \in \mathbf{Z}$, $mx = m$," and statement ♣ is "$x = 1$." Again, the setup of our proof will be this: assume ♡ is true; then try to show that ♣ follows.

Proof of Proposition 1.19. We assume (in addition to what we already know from previous propositions and the axioms) that somebody gives us an $x \in \mathbf{Z}$ and the information that there is some nonzero $m \in \mathbf{Z}$ for which $mx = m$. We first use Axiom 1.3:

$$m \cdot x = m = m \cdot 1\,,$$

and then apply Axiom 1.5 to the left- and right-hand sides of this last equation (note that $m \neq 0$) to deduce that $x = 1$. In summary, assuming x has the property that $mx = m$ for some nonzero $m \in \mathbf{Z}$, we conclude that $x = 1$, and this proves our if–then statement. □

Here are some more propositions about inverses and cancellation:

Proposition 1.20. *For all $m, n \in \mathbf{Z}$, $(-m)(-n) = mn$.*

Proof. Let $m, n \in \mathbf{Z}$. By Axiom 1.4,

$$m + (-m) = 0 \qquad \text{and} \qquad n + (-n) = 0\,.$$

Multiplying both sides of the first equation (on the right) by n and the second equation (on the left) by $-m$ gives, after applying Proposition 1.14 on the right-hand sides,

$$(m + (-m))\,n = 0 \qquad \text{and} \qquad (-m)\,(n + (-n)) = 0\,.$$

With Axiom 1.1(iii) and Proposition 1.6 we deduce

$$mn + (-m)n = 0 \qquad \text{and} \qquad (-m)n + (-m)(-n) = 0\,.$$

It remains to use Axiom 1.1(i) on the left and then Proposition 1.10 to conclude

$$mn = (-m)(-n). \qquad \square$$

Corollary 1.21. $(-1)(-1) = 1$.

The word corollary is used for a statement that is an straightforward consequence of the previous proposition.

Proposition 1.22.

(i) *For all* $m \in \mathbf{Z}$, $-(-m) = m$.

(ii) $-0 = 0$.

Proposition 1.23. *Given* $m, n \in \mathbf{Z}$ *there exists one and only one* $x \in \mathbf{Z}$ *such that* $m + x = n$.

Later (once we have introduced subtraction) we will call this solution $n - m$.

This proposition is an *existence and uniqueness* statement, expressed by the phrase **one and only one**. For given integers m and n, it says that a solution, x, of the equation $m + x = n$ exists (this is the existence part), and that if there appear to be two solutions they must be equal (the uniqueness part).

The word unique has heavy connotations in ordinary speech. In mathematics uniqueness simply means that if they both fit they must be equal.

Proof of Proposition 1.23. The integer $x = (-m) + n$ is a solution, since

$$m + ((-m) + n) = (m + (-m)) + n = 0 + n = n$$

(here we have used Axioms 1.1 and 1.4, and Proposition 1.7).

To prove uniqueness, assume x_1 and x_2 are both solutions to $m + x = n$, i.e.,

$$m + x_1 = n \qquad \text{and} \qquad m + x_2 = n.$$

Since the right-hand sides are equal, we can equate the left-hand sides to deduce

$$m + x_1 = m + x_2,$$

and Proposition 1.9 implies that $x_1 = x_2$. $\square$

Proposition 1.24. *Let* $x \in \mathbf{Z}$. *If* $x \cdot x = x$ *then* $x = 0$ *or* 1.

Proposition 1.25. *For all* $m, n \in \mathbf{Z}$:

(i) $-(m + n) = (-m) + (-n)$.

(ii) $-m = (-1)m$.

(iii) $(-m)n = m(-n) = -(mn)$.

Proposition 1.26. *Let* $m, n \in \mathbf{Z}$. *If* $mn = 0$, *then* $m = 0$ *or* $n = 0$.

Propositions 1.24 and 1.26 contain the innocent-looking word **or**. In everyday language, the meaning of "or" is not always clear. It can mean an *exclusive or* (as in "either ... or ... but not both") or an *inclusive or* (as in "either ... or ... or both"). In mathematics, the word "or," without further qualification, is always inclusive. For example, in Proposition 1.26 it might well happen that both m and n are zero.

This is so important that we will say it again: In mathematics, "$\heartsuit$ or $\clubsuit$" always means either $\heartsuit$, or $\clubsuit$, or both $\heartsuit$ and $\clubsuit$.

Proof of Proposition 1.26. Again we have an if–then statement, so we assume that the integers m and n satisfy $mn = 0$. We need to prove that either $m = 0$ or $n = 0$ (or both). One idea you might have is to rewrite 0 on the right-hand side of the equation $mn = 0$ as $m \cdot 0$ (using Proposition 1.14):

$$m \cdot n = m \cdot 0 . \tag{1.1}$$

This new equation suggests that we use Axiom 1.5 to cancel m on both sides. We have to be careful here: we can do that only if $m \neq 0$. But that is no problem: if $m = 0$ we are done, since then the statement "$m = 0$ or $n = 0$" is true (note that in that case it might still happen that $n = 0$). If $m \neq 0$, we cancel m in (1.1) to deduce $n = 0$, which again means that the statement "$m = 0$ or $n = 0$" holds. In summary, we have shown that if $mn = 0$ then $m = 0$ or $n = 0$. □

In a particular proof, it might be advantageous to switch the roles of $\heartsuit$ and $\clubsuit$ (which you may do freely, since the statement "$\heartsuit$ or $\clubsuit$" is symmetric in $\heartsuit$ and $\clubsuit$).

Our proof illustrates how to approach an "or" statement: if our goal is to prove "$\heartsuit$ or $\clubsuit$" it suffices to prove *one* of $\heartsuit$ and $\clubsuit$. In our proof, $\heartsuit$ was the statement "$m = 0$" and we really needed to worry only about the case that $\heartsuit$ is false and then we needed to prove that $\clubsuit$ is true.

In contrast, when we need to prove an "and" statement, we must prove *two* statements.

Here is something you may try to show: Assuming Axioms 1.1–1.5 we proved Proposition 1.26. On the other hand, if we assume Axioms 1.1–1.4 and the statement of Proposition 1.26, we can prove the statement of Axiom 1.5. In other words, we could have taken Proposition 1.26 as an axiom in place of Axiom 1.5.

1.3 Subtraction

By now you have probably become accustomed to the fact that we use boldface to define a term (such as subtraction in this case).

We now define a new binary operation on **Z**, called $-$ and known as **subtraction**:

$$m - n \qquad \text{is defined to be} \qquad m + (-n) .$$

Proposition 1.27. *For all $m, n, p, q \in \mathbf{Z}$:*

(i) $(m - n) + (p - q) = (m + p) - (n + q)$.

(ii) $(m-n)-(p-q)=(m+q)-(n+p)$.

(iii) $(m-n)(p-q)=(mp+nq)-(mq+np)$.

(iv) $m-n=p-q$ *if and only if* $m+q=n+p$.

(v) $(m-n)p=mp-np$.

The phrase "if and only if" refers to two if–then statements: "♡ if and only if ♣" means: if ♡ then ♣ and if ♣ then ♡. We will say more about this in Section 3.2.

Proof of (i).

$$
\begin{aligned}
(m-n)+(p-q) &\overset{\text{def.}}{=} (m+(-n))+(p+(-q))\\
&\overset{\text{Prop. 1.11(ii)}}{=} ((m+(-n))+p)+(-q)\\
&\overset{\text{Ax. 1.1(ii)}}{=} (m+((-n)+p))+(-q)\\
&\overset{\text{Ax. 1.1(i)}}{=} (m+(p+(-n)))+(-q)\\
&\overset{\text{Ax. 1.1(ii)}}{=} ((m+p)+(-n))+(-q)\\
&\overset{\text{Prop. 1.11(ii)}}{=} (m+p)+((-n)+(-q))\\
&\overset{\text{Prop. 1.25(i)}}{=} (m+p)+(-(n+q))\\
&\overset{\text{def.}}{=} (m+p)-(n+q)
\end{aligned}
$$
□

Take a look at Axiom 1.1(i): for all $m,n \in \mathbf{Z}$, $m+n=n+m$. In words, this says that the binary operation $+$ is commutative. Notice that $-$ is a binary operation that is not commutative. For example, $1-0 \neq 0-1$.

What to say and what to omit. Here, again, we have gone the route of saying exactly what axioms and propositions we are using. Try rewriting this proof saying less, but still doing it in a manner that would convince both you and the reader that you know what you are doing. Discuss this with your instructor in class.

1.4 Philosophical Questions

This chapter has been an illustration of the axiomatic method. We avoided philosophical discussions about the integers—questions like "what *is* an integer?" and "in what sense do integers exist?" We simply agreed that, for our purposes, a set satisfying some axioms is assumed to exist. And we explored the consequences. Everything in this book is introduced on that basis. Later, we will need more axioms as we enrich our theory.

We have stated our axioms. But you could well ask, aren't we really assuming other hidden axioms as well? For example, aren't we assuming precise and organized knowledge about what follows logically from what? This is a difficult question. On the one hand, almost all people agree on what constitutes a correct logical deduction. We could call this the *unexamined basis of logic*. Based on unexamined intuition, highly trained scientists and mathematicians, lawyers and business people all seem to agree on what is a correct logical deduction. But we should tell you that there is a vast literature, thousands of years' worth in mathematics and philosophy, trying to get a sharper understanding of logic and its axiomatic basis. This could be called *examined logic*. Almost everyone, educated or uneducated, proceeds without knowledge of examined logic. That is what we are doing in this book. In summary, we are unable to state all the hidden axioms that we are tacitly assuming.

But, you might ask, in that case why bother with the axioms of this chapter? In our minds we are making a distinction between (a) logic/set theory and (b) mathematics that is built on foundations in logic/set theory. We are allowing ourselves (and you) to be intuitive about the first, but we are demanding precision based on axioms for the second. This is a necessary compromise. Without it we would never get past the difficulties of mathematical foundations. Mathematics is rich and applicable, and those difficulties almost never impinge on the mainstream parts of the subject. In particular, by organizing information about the integers axiomatically, we have seen in this chapter that much mathematics can be deduced from a small set of axioms. This principle will become clearer and more dramatic as we proceed.

As much as we stress a logical, axiomatic approach to number systems in this book, we should not forget that there are human beings behind mathematical development. Numbers are ancient and certainly did not develop historically the way we started with **Z**. In addition to the natural numbers (i.e., the positive integers, which we will introduce in the next chapter), positive rational numbers (i.e., fractions, which we will study in Section 11.1) and certain irrational numbers such as π go back millennia (we will discuss irrational numbers in Section 11.2). Negative integers and fractions can be traced back to India in the sixth century. The construction that we employ in this book emerged only in the nineteenth century. For anyone who wishes to find out more about the rich history of number systems, we recommend the lovely book *Numbers*, by H.-D. Ebbinghaus, H. Hermes, F. Hirzebruch, M. Koecher, K. Mainzer, J. Neukirch, A. Prestel, and R. Remmert (Springer, 1995).

Review Questions. Do you understand what an axiom is? Do you understand what it means to say that a proposition is deduced from the axioms?

Weekly reminder: Reading mathematics is not like reading novels or history. You need to think slowly about every sentence. Usually, you will need to reread the same material later, often more than one rereading.

This is a short book. Its core material occupies about 140 pages. Yet it takes a semester for most students to master this material. In summary: read line by line, not page by page.

Chapter 2
Natural Numbers and Induction

Suppose that we think of the integers lined up like dominoes. The inductive step tells us that they are close enough for each domino to knock over the next one, the base case tells us that the first domino falls over, the conclusion is that they all fall over. The fault in this analogy is that it takes time for each domino to fall and so a domino which is a long way along the line won't fall over for a long time. Mathematical implication is outside time.
Peter J. Eccles (*An Introduction to Mathematical Reasoning*, p. 41)

Before You Get Started. From previous mathematics, you are accustomed to the symbol $<$. If we write $7 < 9$ you read it as "7 is less than 9," and if we write $m < n$ you read it as "m is less than n." But what should, for example, "n greater than 0" mean? If you look back over what we have done so far, you will notice that we have not ordered the integers: even the statement $0 < 1$ does not appear. Here we impose another axiom on $\mathbf{Z}$ to handle these questions. This axiom will specify which integers are to be considered positive. What should it mean for an integer to be positive? Try to come up with an axiomatic way to describe positive integers. And then think about how we could use positive integers to define the symbol $<$.

Another question: Is $\mathbf{Z}$ an infinite set? And what does this mean? Without Axiom 2.1 below we have no answer. We deal with this in Chapter 13.

M. Beck and R. Geoghegan, *The Art of Proof: Basic Training for Deeper Mathematics*, Undergraduate Texts in Mathematics, DOI 10.1007/978-1-4419-7023-7_2,

2.1 Natural Numbers

You may have noticed that while we have defined and studied the integers in the previous chapter, nothing in that chapter allows us to say which integers are "positive." To deal with this we assume another axiom:

We write $A \subseteq B$ ("A is a subset of B") when every member of the set A is a member of B, i.e.: if $x \in A$ then $x \in B$. We discuss this in detail in Section 5.1.

Axiom 2.1. *There exists a subset $\mathbf{N} \subseteq \mathbf{Z}$ with the following properties:*

(i) *If $m, n \in \mathbf{N}$ then $m + n \in \mathbf{N}$.*

(ii) *If $m, n \in \mathbf{N}$ then $mn \in \mathbf{N}$.*

(iii) $0 \notin \mathbf{N}$.

(iv) *For every $m \in \mathbf{Z}$, we have $m \in \mathbf{N}$ or $m = 0$ or $-m \in \mathbf{N}$.*

People use the word negative in two different ways: for the additive inverse of a number, and in the sense of our definition here. For example, $-(-3)$, the negative of -3, is positive. Be on the watch for confusion arising from this.

We call the members of $\mathbf{N}$ **natural numbers** or **positive integers**. A **negative integer** is an integer that is not positive and not zero.

Proposition 2.2. *For $m \in \mathbf{Z}$, one and only one of the following is true: $m \in \mathbf{N}$, $-m \in \mathbf{N}$, $m = 0$.*

Proof. Axiom 2.1(iv) tells us that for each $m \in \mathbf{Z}$, at least one of the three statements $m \in \mathbf{N}$, $-m \in \mathbf{N}$, $m = 0$ is true. The hard part is to show that *only one* of the three statements applies. If $m = 0$, Proposition 1.22 says that $-m = -0 = 0$, so by Axiom 2.1(iii), $m \notin \mathbf{N}$ and $-m \notin \mathbf{N}$.

Now it remains to prove that if $m \neq 0$, then m and $-m$ cannot both be in $\mathbf{N}$. We use a technique called **proof by contradiction**. The idea is simple: Say we want to prove that some statement $\heartsuit$ is true. Then we start our argument by supposing that $\heartsuit$ is *false* and we show that this leads to a contradiction. Typically we deduce from the falsity of $\heartsuit$ some other statement that is obviously false, such as $0 = 1$ or some other negation of one of our axioms.

To prove Proposition 2.2 we need to show that, given an $m \neq 0$, m and $-m$ are not both in $\mathbf{N}$. The negation of this conclusion is the statement that both m and $-m$ are in $\mathbf{N}$. So we suppose (hoping to arrive at a contradiction) that m and $-m$ are both in $\mathbf{N}$. Axiom 2.1(i) then tells us that

$$m + (-m) \in \mathbf{N}.$$

But we also know, by Axiom 1.4, that

$$m + (-m) = 0.$$

Combining these two statements yields $0 \in \mathbf{N}$. But this contradicts Axiom 2.1(iii)—the statement $0 \in \mathbf{N}$ is precisely the negation of that axiom. This contradiction means that our assumption that both m and $-m$ are in $\mathbf{N}$ must be false, that is, at most one of m and $-m$ is in $\mathbf{N}$. □

This is a good place to say more about proof by contradiction. Let ♡ be a mathematical statement. Then "not ♡" is also a mathematical statement. To say "♡ is false" is the same thing as saying "(not ♡) is true." In other words, we assume the **Law of the Excluded Middle**, that ♡ must be either true or false—there is no middle ground. This assumption in our logic lies behind proof by contradiction. If you want to prove that ♡ is true you can do so directly, or you can prove it by contradiction:

Template for Proof by Contradiction. You wish to prove that ♡ is true. Suppose (by way of contradiction) that ♡ is false. Deduce from this, in as many steps as is necessary, that ♠ is true, where ♠ is a statement that you know to be false. Conclude that ♡ must be true, because the supposition that ♡ is false led you to a contradiction.

Proof by contradiction is useful for statements that begin "There does not exist . . . " (Suppose it did exist . . .)

The validity of proof by contradiction depends on there being no contradictions built into our axioms. It is the authors' job to make sure our axioms lead to no (known) contradictions.

To be honest we should tell you more: you probably believe (as we do) that the axioms we have introduced so far do not contradict each other, but an amazing theorem of Kurt Gödel (1906–1978) says that this cannot be proved: one cannot prove that a given system of axioms is consistent without moving to a "higher" theory—roughly speaking, one with more axioms.

In the late nineteenth and early twentieth century there was controversy in the mathematical world as to whether a theorem is really proved if it is only proved by contradiction. There was a feeling that a proof is stronger and more convincing if it is not by contradiction. With the rise of computer science and interest in computability this has become a serious issue in certain circles. We can say, however, that today proof by contradiction is accepted as valid by all but a tiny number of mathematicians.

Here is another proposition that you might try to prove by contradiction:

Proposition 2.3. $1 \in \mathbf{N}$.

2.2 Ordering the Integers

Now that we have introduced $\mathbf{N}$, we can order the integers. Here is how we do it:

Let $m, n \in \mathbf{Z}$. The statements $m < n$ (m **is less than** n) and $n > m$ (n **is greater than** m) both mean that

$$n - m \in \mathbf{N}.$$

The notations $m \leq n$ (m **is less than or equal to** n) and $n \geq m$ (n **is greater than or equal to** m) mean that

$$m < n \qquad \text{or} \qquad m = n.$$

Proposition 2.4. *Let $m, n, p \in \mathbf{Z}$. If $m < n$ and $n < p$ then $m < p$.*

Proof. Assume $m < n$ and $n < p$, that is, $n - m \in \mathbf{N}$ and $p - n \in \mathbf{N}$. Then by Axiom 2.1(i),

$$p - m = (p - n) + (n - m) \in \mathbf{N},$$

that is, $m < p$. □

What to say and what to omit. In this proof of Proposition 2.4 we explain how we are using Axiom 2.1, but we do not spell out how we are using the earlier axioms. Do you think what we have written here is sufficiently clear?

Proposition 2.5 implies that $\mathbf{N}$ *is an infinite set; we will discuss this in detail in Chapter 13 (see Proposition 13.8).*

Proposition 2.5. *For each* $n \in \mathbf{N}$ *there exists* $m \in \mathbf{N}$ *such that* $m > n$.

Proposition 2.6. *Let* $m, n \in \mathbf{Z}$. *If* $m \leq n \leq m$ *then* $m = n$.

Proposition 2.7. *Let* $m, n, p, q \in \mathbf{Z}$.

(i) *If* $m < n$ *then* $m + p < n + p$.

(ii) *If* $m < n$ *and* $p < q$ *then* $m + p < n + q$.

(iii) *If* $0 < m < n$ *and* $0 < p \leq q$ *then* $mp < nq$.

(iv) *If* $m < n$ *and* $p < 0$ *then* $np < mp$.

Proof of (iii). Assume $0 < m < n$ and $0 < p \leq q$, i.e.,

Why are the statements $m > 0$ *and* $m \in \mathbf{N}$ *equivalent?*

$$m \in \mathbf{N},$$
$$n - m \in \mathbf{N},$$
$$p \in \mathbf{N}, \text{ and}$$
$$q - p \in \mathbf{N} \text{ or } p = q.$$

We need to prove

$$mp < nq, \qquad \text{i.e.,} \qquad nq - mp \in \mathbf{N}.$$

It is sometimes useful to separate an argument into cases, but you are not obliged to do this.

Case 1: $p = q$.

$$nq - mp = np - mp = (n - m)p \in \mathbf{N}$$

by Proposition 1.27(v) and Axiom 2.1(ii), because both $n - m \in \mathbf{N}$ and $p \in \mathbf{N}$.

Case 2: $p \neq q$.

We have $q - p \in \mathbf{N}$, and so

$$nq - mp = (nq - mq) + (mq - mp) = (n - m)q + m(q - p) \in \mathbf{N},$$

by Proposition 2.4 (from which we conclude $q > 0$, i.e., $q \in \mathbf{N}$) and Axiom 2.1(i) and (ii). □

Proposition 2.8. *Let* $m, n \in \mathbf{Z}$. *Exactly one of the following is true:* $m < n$, $m = n$, $m > n$.

Proposition 2.9. *Let $m \in \mathbf{Z}$. If $m \neq 0$ then $m^2 \in \mathbf{N}$.*

Here m^2 means $m \cdot m$.

Proposition 2.10. *The equation $x^2 = -1$ has no solution in $\mathbf{Z}$.*

Proposition 2.10 is an opportunity for a proof by contradiction.

Proposition 2.11. *Let $m \in \mathbf{N}$ and $n \in \mathbf{Z}$. If $mn \in \mathbf{N}$, then $n \in \mathbf{N}$.*

Proposition 2.12. *For all $m, n, p \in \mathbf{Z}$:*

(i) *$-m < -n$ if and only if $m > n$.*

(ii) *If $p > 0$ and $mp < np$ then $m < n$.*

(iii) *If $p < 0$ and $mp < np$ then $n < m$.*

(iv) *If $m \leq n$ and $0 \leq p$ then $mp \leq np$.*

Two sets A and B are **equal** (in symbols, $A = B$) if

We will study equality of sets in detail in Section 5.1.

$$A \subseteq B \qquad \text{and} \qquad B \subseteq A.$$

If you recall that $A \subseteq B$ means

$$\text{if } x \in A \text{ then } x \in B$$

and $B \subseteq A$ means

$$\text{if } x \in B \text{ then } x \in A,$$

we can define the set equality $A = B$ also as follows:

$$x \in A \text{ if and only if } x \in B.$$

Here is an example:

Proposition 2.13. $\mathbf{N} = \{n \in \mathbf{Z} : n > 0\}$.

The notation $S = \{x \in A : x \text{ satisfies } \clubsuit\}$ means that the elements of the set S are precisely those $x \in A$ that satisfy statement $\clubsuit$. An alternative notation is $\{x \in A \mid x \text{ satisfies } \clubsuit\}$.

Proof. We need to prove $\mathbf{N} \subseteq \{n \in \mathbf{Z} : n > 0\}$ and $\mathbf{N} \supseteq \{n \in \mathbf{Z} : n > 0\}$, that is, for $n \in \mathbf{Z}$ we need to prove

$$n \in \mathbf{N} \qquad \text{if and only if} \qquad n \in \mathbf{Z} \text{ and } n > 0.$$

But since $n - 0 = n$, this is precisely the definition of the statement $n > 0$. □

2.3 Induction

The introduction of **N** allowed us to discuss the notions of "positive integer" and "$m < n$." Mathematics also uses **N** in a different way: to prove theorems by a method called induction. To admit this new method of proof into our mathematics, we introduce another axiom. We begin with a proposition.

Part (i) is just a repetition of Proposition 2.3.

Proposition 2.14.

(i) $1 \in \mathbf{N}$.

(ii) *If* $n \in \mathbf{N}$ *then* $n+1 \in \mathbf{N}$.

Our new axiom says that $\mathbf{N}$ is the smallest subset of $\mathbf{Z}$ that satisfies Proposition 2.14.

Axiom 2.15 (Induction Axiom). *If a subset* $A \subseteq \mathbf{Z}$ *satisfies*

(i) $1 \in A$ *and*

(ii) *if* $n \in A$ *then* $n+1 \in A$,

then $\mathbf{N} \subseteq A$.

Our aim in this section is to explain how this axiom is used.

Proposition 2.16. *Let* $B \subseteq \mathbf{N}$ *be such that:*

(i) $1 \in B$ *and*

(ii) *if* $n \in B$ *then* $n+1 \in B$.

Then $B = \mathbf{N}$.

Proof. The hypothesis says that $B \subseteq \mathbf{N}$. By Axiom 2.15, $\mathbf{N} \subseteq B$. Therefore $B = \mathbf{N}$. □

Proposition 2.16 gives us the new method of proof:

Theorem 2.17 (Principle of mathematical induction—first form). *Let* $P(k)$ *be a statement depending on a variable* $k \in \mathbf{N}$. *In order to prove the statement "*$P(k)$ *is true for all* $k \in \mathbf{N}$*" it is sufficient to prove:*

(i) $P(1)$ *is true and*

(ii) *for any given* $n \in \mathbf{N}$, *if* $P(n)$ *is true then* $P(n+1)$ *is true.*

The notation $S := \heartsuit$ *means we are defining* S *to be* $\heartsuit$.

Proof. Let

$$B := \{k \in \mathbf{N} : P(k) \text{ is true}\},$$

and assume that

(i) we can prove $P(1)$ is true and

(ii) if $P(n)$ is true then $P(n+1)$ is true.

This means $1 \in B$; and if $n \in B$ then $n+1 \in B$. By Proposition 2.16, $B = \mathbf{N}$, in other words, $P(k)$ is true for all $k \in \mathbf{N}$. □

So far, the only integers with names are 0 and 1. We name some more:

We will use the symbol	2	to denote 1 + 1
	3	2 + 1
	4	3 + 1
	5	4 + 1
	6	5 + 1
	7	6 + 1
	8	7 + 1
	9	8 + 1.

The integers 0, 1, 2, 3, 4, 5, 6, 7, 8, 9 are called **digits**. All except 0 belong to **N**, i.e., are natural numbers; this follows from Proposition 2.14. From previous experience, you know what we mean by symbols like 63, 721, −2719; they are names of other integers. While we may use these names informally (for example, to number the pages of this book) we will give a proper treatment of the "base 10" system in Chapter 7.

Proofs that use Theorem 2.17 are called *proofs by induction*. Here are some examples:

Proposition 2.18.

(i) *For all* $k \in \mathbf{N}$, $k^3 + 2k$ *is divisible by* 3.

(ii) *For all* $k \in \mathbf{N}$, $k^4 - 6k^3 + 11k^2 - 6k$ *is divisible by* 4.

(iii) *For all* $k \in \mathbf{N}$, $k^3 + 5k$ *is divisible by* 6.

Here k^3 means $k^2 \cdot k$, and k^4 means $k^3 \cdot k$.

Proof of (i). We will use induction on k. Let $P(k)$ denote the statement

$$k^3 + 2k \text{ is divisible by } 3\,.$$

The induction principle states that we first need to check $P(1)$—the **base case**. That is, we must check the statement "$1^3 + 2 \cdot 1$ is divisible by 3," which is certainly true: $1^3 + 2 \cdot 1 = 1 + 2 = 3$, and 3 is divisible by 3.

Next comes the **induction step**, that is, we assume that $P(n)$ is true for some $n \in \mathbf{N}$ and show that $P(n+1)$ holds as well. So assume that $n^3 + 2n$ is divisible by 3, that is, there exists $y \in \mathbf{Z}$ such that

$$n^3 + 2n = 3y\,.$$

Our goal is to show that $(n+1)^3 + 2(n+1)$ is divisible by 3, that is, we need to show the existence of $z \in \mathbf{Z}$ such that

$$(n+1)^3+2(n+1)=3z.$$

But the left-hand side of this equation can be rewritten as

$$\begin{aligned}(n+1)^3+2(n+1) &= n^3+3n^2+3n+1+2n+2\\ &= \left(n^3+2n\right)+3n^2+3n+3\\ &= 3y+3n^2+3n+3\\ &= 3\left(y+n^2+n+1\right).\end{aligned}$$

So we can set $z=y+n^2+n+1$, which is an integer (because y and n are in $\mathbf{Z}$). Thus we have proved that there exists $z\in\mathbf{Z}$ such that $(n+1)^3+2(n+1)=3z$, which concludes our induction step. □

A ladder disappears into the clouds, meaning that the ladder does not have a top rung. For every $k\in\mathbf{N}$ this ladder has a kth rung. You want to persuade a six-year old girl that she can climb to any desired rung; she has unlimited time and energy. You break the issue down:

- *"Can't you climb onto the first rung?" "Yes." (This is the base case.)*
- *"If I place you on the nth rung, can't you climb to the $(n+1)^{th}$ rung?" "Yes." (This is the induction step.)*
- *"Doesn't it follow that (with enough time and energy) you can climb to the kth rung for any k?" "Yes."*

Note that the ladder has no top. The principle of induction does not imply that the child can climb infinitely many rungs. Rather, she can climb to any rung.

What to say and what to omit. This proof of Proposition 2.18(i) would be regarded by most instructors as sufficiently detailed. Dependence on the axioms is not spelled out, but the tone and style suggest that the writer understands what is going on, that (s)he could fill in whatever details have been omitted if challenged, and that the writer believes the key ideas have been conveyed to the (qualified) reader. But what if the reader does not understand how the steps in the proof are justified? Would you be able to spell out missing details when challenged? Test yourself.

Template for Proofs By Induction.
Formulate $P(k)$.
Base case: prove that $P(1)$ is true.
Induction step: Let $n\in\mathbf{N}$; assume that $P(n)$ is true. Then, using the assumption that $P(n)$ is true, prove that $P(n+1)$ is also true.

Project 2.19. Come up with (and prove) other divisibility statements.

Proposition 2.20. *For all $k\in\mathbf{N}$, $k\geq 1$.*

Proposition 2.21 is another opportunity to construct a proof by contradiction.

Proposition 2.21. *There exists no integer x such that $0<x<1$.*

Corollary 2.22. *Let $n\in\mathbf{Z}$. There exists no integer x such that $n<x<n+1$.*

Proposition 2.23. *Let $m,n\in\mathbf{N}$. If n is divisible by m then $m\leq n$.*

Proposition 2.24. *For all $k\in\mathbf{N}$, $k^2+1>k$.*

It is sometimes useful to start inductions at an integer other than 1:

Theorem 2.25 (Principle of mathematical induction—first form revisited). *Let $P(k)$ be a statement, depending on a variable $k\in\mathbf{Z}$, that makes sense for all $k\geq m$, where m is a fixed integer. In order to prove the statement "$P(k)$ is true for all $k\geq m$" it is sufficient to prove:*

(i) *$P(m)$ is true and*

(ii) *for any given $n \geq m$, if $P(n)$ is true then $P(n+1)$ is true.*

Proof. Let $Q(k)$ be the statement $P(k+m-1)$, that is, $Q(1) = P(m)$, $Q(2) = P(m+1)$, etc. The original Theorem 2.17 on induction states that to prove the statement "$Q(k)$ is true for all $k \geq 1$" it is sufficient to prove that $Q(1)$ is true and for $n \geq 1$, if $Q(n)$ is true then $Q(n+1)$ is true. But this is equivalent to saying that to prove the statement "$P(k)$ is true for all $k \geq m$" it is sufficient to prove that $P(m)$ is true and for $n \geq m$, if $P(n)$ is true then $P(n+1)$is true. □

Here is an example where Theorem 2.25 is useful:

Proposition 2.26. *For all integers $k \geq -3$, $3k^2 + 21k + 37 \geq 0$.*

Proof. We use induction on k. The base case ($k = -3$) holds because

$$3(-3)^2 + 21(-3) + 37 = 1 \geq 0\,.$$

For the induction, assume that $3n^2 + 21n + 37 \geq 0$ for some $n \geq -3$. Then

$$3(n+1)^2 + 21(n+1) + 37 = 3n^2 + 27n + 61 \geq 6n + 24\,,$$

by the induction hypothesis. Because $n \geq -3$, $6n + 24 \geq 6 \geq 0$, and thus

$$3(n+1)^2 + 21(n+1) + 37 \geq 0\,,$$

which completes our induction step. □

Proposition 2.27. *For all integers $k \geq 2$, $k^2 < k^3$.*

One can also prove Proposition 2.27 without induction—try both ways.

Project 2.28. Determine for which natural numbers $k^2 - 3k \geq 4$ and prove your answer.

2.4 The Well-Ordering Principle

Let $A \subseteq \mathbf{Z}$ be nonempty. If there exists $b \in A$ such that for all $a \in A$, $b \leq a$, then b is the **smallest element** of A, in which case we write $b = \min(A)$.

Our language suggests that $\min(A)$, if it exists, is unique, which you should prove.

Example 2.29. By Propositions 2.3 and 2.20, 1 is the smallest element of $\mathbf{N}$.

Example 2.30. The set of integers divisible by 6 does not have a smallest element. On the other hand, the set of positive integers divisible by 6 does have a smallest element: what is it?

Project 2.31. What is the smallest element of

$$\{3m+8n : m,n \in \mathbf{Z},\, m \geq n \geq -4\}?$$

Theorem 2.32 (Well-Ordering Principle). *Every nonempty subset of* $\mathbf{N}$ *has a smallest element.*

Here we show that Theorem 2.32 follows from Axiom 2.15. One can also deduce Axiom 2.15 from Theorem 2.32—in other words, we could have stated Theorem 2.32 as an axiom, and then proved the statement of Axiom 2.15 as a theorem.

Theorem 2.32 is well worth memorizing; you will see it in action several times in later chapters. We will give a first application after the proof, which is a somewhat subtle application of the principle of induction.

Proof. Consider the set

$$N := \left\{ k \in \mathbf{N} : \begin{array}{l} \text{every subset of } \mathbf{N} \text{ that contains an} \\ \text{integer} \leq k \text{ has a smallest element} \end{array} \right\}.$$

Our goal is to prove that $N = \mathbf{N}$. We will prove by induction that every natural number is in N.

Base case: In Example 2.29 we saw that 1 is the smallest element of $\mathbf{N}$, so if a subset of $\mathbf{N}$ contains 1, then 1 is its smallest element. This shows $1 \in N$.

For the induction step, assume that $n \in N$, that is, every subset of $\mathbf{N}$ that contains an integer $\leq n$ has a smallest element. Now let S be a subset of $\mathbf{N}$ that contains an integer $\leq n+1$. We need to prove that S also has a smallest element. If S contains an integer $\leq n$, then S has a smallest element by the induction hypothesis. Otherwise (i.e., when S does not contain an integer $\leq n$), S must contain $n+1$, and this integer is the smallest element of S. □

Proposition 2.33. *Let* A *be a nonempty subset of* $\mathbf{Z}$ *and* $b \in \mathbf{Z}$*, such that for each* $a \in A$, $b \leq a$. *Then* A *has a smallest element.*

We now give a first application of the Well-Ordering Principle. Given two integers m and n, we define the number $\gcd(m,n)$ to be the smallest element of

$$S := \{k \in \mathbf{N} : k = mx + ny \text{ for some } x,y \in \mathbf{Z}\}.$$

(S is empty when $m = n = 0$, in which case we define $\gcd(0,0) := 0$.) We use the Well-Ordering Principle (Theorem 2.32) here to ensure that the definition of $\gcd(m,n)$ makes sense: Without Theorem 2.32 it is not clear that S *has* a smallest element. However, to apply Theorem 2.32, we should make sure that S is not empty:

Proposition 2.34. *If* m *and* n *are integers that are not both* 0*, then*

$$S = \{k \in \mathbf{N} : k = mx + ny \text{ for some } x,y \in \mathbf{Z}\}$$

is not empty.

Project 2.35. Compute $\gcd(4,6)$, $\gcd(7,13)$, $\gcd(-4,10)$, and $\gcd(-5,-15)$.

The notation gcd *stands for* greatest common divisor, *and Projects 2.35 and 2.36 should convince you that our definition gives what you thought to be the greatest common divisor of two integers. We will discuss this further in Section 6.4.*

Project 2.36. Given a nonzero integer n, compute $\gcd(0,n)$ and $\gcd(1,n)$.

Review Question. Are you able to use the method of induction correctly?

Weekly reminder: Reading mathematics is not like reading novels or history. You need to think slowly about every sentence. Usually, you will need to reread the same material later, often more than one rereading.

This is a short book. Its core material occupies about 140 pages. Yet it takes a semester for most students to master this material. In summary: read line by line, not page by page.

Chapter 3
Some Points of Logic

Noel sing we, both all and some.
A line in a fifteenth century English Christmas carol

DILBERT: © Scott Adams / Dist. by United Feature Syndicate, Inc. Reprinted with permission.

Before You Get Started. We have already used phrases like *for all*, *there exists*, *and*, *or*, *if... then*, etc. What exactly do these mean? How would the meaning of a statement such as Axiom 1.4 change if we switched the order of some of these phrases? We will also need to express the *negations* of mathematical statements. Think about what the negation of an *and* statement should look like; for example, what are the negations of the statements "John and Mary like cookies" or "there exist positive integers that are not prime"?

M. Beck and R. Geoghegan, *The Art of Proof: Basic Training for Deeper Mathematics*, Undergraduate Texts in Mathematics, DOI 10.1007/978-1-4419-7023-7_3,

3.1 Quantifiers

The words for any *usually means* $\forall$*, but* for any *is sometimes used (misused in our opinion) to mean* $\exists$.

The symbol $\forall$ means *for all* or *for each* or *for every* or or *whenever*. Whether or not you think these five phrases mean the same thing in ordinary conversation, they do mean the same thing in mathematics.

The symbol $\exists$ means *there exists* or (in the plural) *there exist*. It is always qualified by a property: i.e., one usually says "$\exists$ ... such that ..." Another translation of $\exists$ is *for some*. Here are some examples.

(i) Axiom 1.2 could be written: $\exists 0 \in \mathbf{Z}$ such that $\forall m \in \mathbf{Z}$, $m+0=m$.

(ii) Axiom 1.4 could be written: $\forall m \in \mathbf{Z}\ \exists n \in \mathbf{Z}$ such that $m+n=0$.

The symbol $\forall$ is the **universal quantifier** and the symbol $\exists$ is the **existential quantifier**. It is instructive to break up the two sentences in the examples:

(i) ($\exists 0 \in \mathbf{Z}$ such that) ($\forall m \in \mathbf{Z}$) $m+0=m$.

(ii) ($\forall m \in \mathbf{Z}$) ($\exists n \in \mathbf{Z}$ such that) $m+n=0$.

Here are some features to note:

- Both statements consist of **quantified segments** of the form ($\exists$ ... such that) and ($\forall$ ...) in a particular order, and then a **final statement**. For example, $m+0=m$ is the final statement in (i), and $m+n=0$ is the final statement in (ii).
- The order is important. In (ii) n depends on m.
- The informal phrase *for some* really means "$\exists$ ... such that"; for example, informally we might write Axiom 1.2 as:

The clumsiness of this sentence should explain why for some *is not a good phrase for beginners to use.*

$$\text{for some } 0 \in \mathbf{Z}, \text{ for all } m \in \mathbf{Z}, \ m+0=m.$$

For a more complicated example, we look again at (ii) above (i.e., at Axiom 1.4) and at the statement that we get by switching the two quantifiers:

(a) ($\forall m \in \mathbf{Z}$) ($\exists n \in \mathbf{Z}$ such that) $m+n=0$.

(b) ($\exists n \in \mathbf{Z}$ such that) ($\forall m \in \mathbf{Z}$) $m+n=0$.

The key fact to note is that in (a) n depends on m; change m and you expect you will have to change n accordingly. This is because the segment involving n comes after the segment involving m. Axiom 1.4 asserts the existence of the additive inverse for a given number m; it is indeed the case that different m's have different additive inverses. So, in words, (a) is best read as: for each $m \in \mathbf{Z}$ there exists $n \in \mathbf{Z}$ such that $m+n=0$.

Now look at (b). This statement asserts the existence of an n that will work as the "additive inverse" for all m—a statement that you know is wrong. Thus

$$(\forall m \in \mathbf{Z})\,(\exists n \in \mathbf{Z} \text{ such that})\ \ldots$$

and

$$(\exists n \in \mathbf{Z} \text{ such that})\ (\forall m \in \mathbf{Z})\ \ldots$$

have quite different meanings. While in our example, one is true and the other is false, the point we are making here is that interchanging the quantified phrases changes the meaning.

You can have several $\forall$-phrases in a row, but they can always be reorganized into one:

$$(\forall \heartsuit)\,(\forall \clubsuit) \qquad \text{has the same meaning as} \qquad (\forall \heartsuit \text{ and } \clubsuit).$$

You can also find yourself saying

$$(\exists \heartsuit \text{ and } \clubsuit \text{ such that})\ \ldots\,.$$

Then you are asserting the existence of two things that in combination have some property ....

Project 3.1. Express each of the following statements using quantifiers.

(i) There exists a smallest natural number.

(ii) There exists no smallest integer.

(iii) Every integer is the product of two integers.

(iv) The equation $x^2 - 2y^2 = 3$ has an integer solution.

Project 3.2. In each of the following cases explain what is meant by the statement and decide whether it is true or false.

(i) For each $x \in \mathbf{Z}$ there exists $y \in \mathbf{Z}$ such that $x + y = 1$.

(ii) There exists $y \in \mathbf{Z}$ such that for each $x \in \mathbf{Z}$, $x + y = 1$.

(iii) For each $x \in \mathbf{Z}$ there exists $y \in \mathbf{Z}$ such that $xy = x$.

(iv) There exists $y \in \mathbf{Z}$ such that for each $x \in \mathbf{Z}$, $xy = x$.

By now you might have guessed that a *for all* statement can be rewritten as an *if then* statement. For example, the statement

$$\text{for all } m \in \mathbf{N},\ m \in \mathbf{Z} \qquad \text{is equivalent to} \qquad \text{if } m \in \mathbf{N} \text{ then } m \in \mathbf{Z}.$$

Uniqueness. The notation $(\exists!\, n \in \mathbf{Z}$ such that $\ldots)$ means that there exists a *unique* $n \in \mathbf{Z}$ with the given property. There are two statements here:

(i) existence ($\exists n \in \mathbf{Z}$ such that ...)

and

(ii) uniqueness (if $n_1 \in \mathbf{Z}$ and $n_2 \in \mathbf{Z}$ both have the given property, then $n_1 = n_2$).

We have seen instances of this concept in Propositions 1.10 and 1.23.

3.2 Implications

Here are a few other commonly used statements for $\heartsuit \Rightarrow \clubsuit$: $\heartsuit$ implies $\clubsuit$. $\heartsuit$ only if $\clubsuit$. $\clubsuit$ if $\heartsuit$. $\clubsuit$ whenever $\heartsuit$. $\heartsuit$ is sufficient for $\clubsuit$. $\clubsuit$ is necessary for $\heartsuit$.

$\heartsuit \Leftrightarrow \clubsuit$ is often expressed as "$\heartsuit$ and $\clubsuit$ are equivalent."

We have already discussed if–then statements. The statement "if $\heartsuit$ then $\clubsuit$" can also be expressed as "$\heartsuit$ implies $\clubsuit$." The symbol $\Rightarrow$ stands for "implies" in this sentence; that is,

$$\heartsuit \Rightarrow \clubsuit \qquad \text{has the same meaning as} \qquad \text{if } \heartsuit \text{ then } \clubsuit.$$

For example, $\heartsuit$ could be the statement "it is raining" and $\clubsuit$ the statement "the street is wet." Then $\heartsuit \Rightarrow \clubsuit$ says "if it is raining, then the street is wet." In the margin we list some other phrases equivalent to $\heartsuit \Rightarrow \clubsuit$. One of them, namely "$\heartsuit$ only if $\clubsuit$," can be confusing and is mostly used in *double implications*: we say "$\heartsuit$ **if and only if** $\clubsuit$" when the two if–then statements $\heartsuit \Rightarrow \clubsuit$ and $\clubsuit \Rightarrow \heartsuit$ are true; notationally we abbreviate this to $\heartsuit \Leftrightarrow \clubsuit$.

Converse. The statement $\clubsuit \Rightarrow \heartsuit$ is called the **converse** of the implication $\heartsuit \Rightarrow \clubsuit$. When you look back at the implications we have encountered so far, you will notice that often *both* $\heartsuit \Rightarrow \clubsuit$ and $\clubsuit \Rightarrow \heartsuit$ are true statements (and then the if-and-only-if statement $\heartsuit \Leftrightarrow \clubsuit$ holds). You have seen if-and-only-if statements a few times: in Propositions 1.27 and 2.12(i), and in the proof of Proposition 2.13. Moreover, several if–then statements you have seen so far can, in fact, be strengthened to if-and-only-if statements. One example is Proposition 2.6: we might as well have stated it as (assuming $m, n \in \mathbf{Z}$)

The statement "if $m = n$ then $m \le n \le m$" is not very enlightening, and so we omitted it in Proposition 2.6.

$$m \le n \le m \qquad \text{if and only if} \qquad m = n.$$

However, in general one has to be careful not to confuse an implication $\heartsuit \Rightarrow \clubsuit$ with its converse $\clubsuit \Rightarrow \heartsuit$: often only one of the two is true. Here is an example of a true implication whose converse is false. Let $m, n \in \mathbf{Z}$.

If $\gcd(m,n) = 3$, then m and n are divisible by 3.

Project 3.3. Construct two more mathematical if–then statements that are true, but whose converses are false.

Contrapositive. An implication $\heartsuit \Rightarrow \clubsuit$ can be rewritten in terms of the negatives of the statements $\heartsuit$ and $\clubsuit$; namely,

$$\heartsuit \Rightarrow \clubsuit \qquad \text{has the same meaning as} \qquad (\text{not } \clubsuit) \Rightarrow (\text{not } \heartsuit).$$

This statement on the right is called the **contrapositive** of $\heartsuit \Rightarrow \clubsuit$. Contrapositives can be useful for proofs: sometimes it is easier to prove (not $\clubsuit$) $\Rightarrow$ (not $\heartsuit$) than to prove $\heartsuit \Rightarrow \clubsuit$.

Example 3.4. Proposition 2.6 says that (assuming $m, n \in \mathbf{Z}$) $m \le n \le m$ implies $m = n$. We could have proved this proposition by showing its contrapositive:

$$\text{if } m \neq n \text{ then } m > n \text{ or } n > m.$$

But this follows immediately from Proposition 2.8.

Here we are using the negation of an and *statement, discussed in Section 3.3. Namely, the negation of $m \le n \le m$ is $m > n$* or *$n > m$.*

Project 3.5. Re-prove some of the if–then propositions in Chapters 1 and 2 by proving their contrapositives.

3.3 Negations

Here we discuss how to negate mathematical statements. We start with two easy cases.

Negation of 'and' statements; negation of 'or' statements. The negation of "$\heartsuit$ and $\clubsuit$" is "(not $\heartsuit$) or (not $\clubsuit$)," and the negation of "$\heartsuit$ or $\clubsuit$" is "(not $\heartsuit$) and (not $\clubsuit$)." Convince yourself that this makes sense. Convince your friends.

Example 3.6. $\exists!$ gives an "and" statement. Thus the negation of a $\exists!$ phrase will be an "or" statement.

Work out the negation of a $\exists!$ phrase in detail.

Negation of if–then statements. What is the negation of the if–then statement $\heartsuit \Rightarrow \clubsuit$? Suppose we tell you, "On Mondays we have lunch at the student union"; mathematically speaking: *if* it is Monday, *then* we have lunch at the student union. You would like to prove us wrong; i.e., you feel that the negation of this statement is true. Then you will probably hang out at the student union on Mondays checking whether we actually show up. In other words, to prove us wrong, you would show us that "today is Monday but you did not have lunch at the union." So the *negation* of the if–then statement ($\heartsuit \Rightarrow \clubsuit$) is the statement ($\heartsuit$ and not $\clubsuit$).

Negation of statements that involve $\forall$ and $\exists$. Once you organize a sentence as quantified phrases followed by a final statement, the negation of that sentence is easily found. Here is the rule:

This is an instance of De Morgan's law. *We will see a version for set unions and intersections in Theorem 5.15.*

(i) Maintaining the order of the quantified segments, change each ($\forall$ ...) segment into a ($\exists$ ... such that) segment;

(ii) change each ($\exists$ ... such that) segment into a ($\forall$ ...) segment;

(iii) negate the final statement.

For example, the statement "Axiom 1.4 does not hold" can be written as

$$(\exists m \in \mathbf{Z} \text{ such that})\ (\forall n \in \mathbf{Z})\ m+n \neq 0.$$

Project 3.7. Negate the following statements.

Note that you do not need to know the meaning of these statements in order to negate them.

(i) Every cubic polynomial has a real root.

(ii) G is normal and H is regular.

(iii) $\exists!\,0$ such that $\forall x,\ x+0=x$.

(iv) The newspaper article was neither accurate nor entertaining.

(v) If $\gcd(m,n)$ is odd, then m or n is odd.

(vi) H/N is a normal subgroup of G/N if and only if H is a normal subgroup of G.

(vii) For each $\varepsilon > 0$ there exists $N \in \mathbf{N}$ such that for all $n \geq N$, $|a_n - L| < \varepsilon$.

3.4 Philosophical Questions

This is a good moment to say more about the words *true* and *false*. In ordinary life, it is often not easy to say that a given statement is true or false; maybe it is neither, maybe it was written down to suggest imprecise ideas. Even deciding what propositions make sense can be difficult. For example:

- She loves me, she loves me not.
- Colorless green ideas sleep furiously.
- What did the professor talk about in class today? Actually it was like totally confusing because he just went on and on about all this stuff about integers, you know, and things like that, and I was like totally not there.

The first of these is poetry, and is expressing a thought entirely different from what the words actually say. The second, a famous example due to the mathematical linguist Noam Chomsky (1928–), is grammatically correct but meaningless. The third—well, what can we say?

We would not consider these to be mathematical statements. It is not easy to say precisely what a mathematical statement is, but this much we can say: it should be a sentence in the ordinary sense, and it should be part of a mathematical discussion. You may not know whether a particular statement is true or false; in fact, much of mathematics is concerned with trying to decide whether a statement is true or false. But it should be clear on first reading that it is the kind of statement that has to be either true or false.

Every time you prove a proposition or theorem, you are showing that the given statement is true.

A statement consists of words, so we should discuss what kinds of words belong in a mathematical statement. It should be the case that we already know the meaning of the words we use in formulating a statement. Therefore we need a dictionary. The custom in this book is that the first time a word is used it is highlighted in boldface. This explanation of the new word in terms of other previously known words is called a **definition**. These definitions are (part of) our dictionary.

Many authors use italics rather than boldface for this; in handwriting you might use underlining.

There is a huge logical problem here: where do we start? How could we possibly write down the first definition? But if you think about it, you will realize that the same problem arises in the learning of languages. If we are learning a second language, we can build up the whole dictionary by using some words from our first language at the beginning. But how did each of us learn his or her first language? That is a deep problem in psychology. What we can all say for sure from our own experience is that we did not learn our first language by building a formal dictionary. Somehow, we knew the meanings of some words and that is what got us started. So, in the same way, we have to start our mathematics by honestly admitting that there are going to be some undefined terms whose meanings we know intuitively. This may seem like a logical mess, but it is real life and we are not able to disentangle it. As a practical matter, the problems discussed here will not cause us trouble.

Review Questions. Do you understand how to use $\forall$ and $\exists$? Do you understand how to write down the negation of a sentence?

Weekly reminder: Reading mathematics is not like reading novels or history. You need to think slowly about every sentence. Usually, you will need to reread the same material later, often more than one rereading.

This is a short book. Its core material occupies about 140 pages. Yet it takes a semester for most students to master this material. In summary: read line by line, not page by page.

Chapter 4
Recursion

Before You Get Started. You have most likely seen sums of the form $\sum_{j=1}^{k} j = 1+2+3+\cdots+k$, or products like $k! = 1\cdot 2\cdot 3\cdots k$. In this chapter we will use the idea behind induction to *define* expressions like these. For example, we can define the sum $1+2+3+\cdots+(k+1)$ by saying, if you know what $1+2+3+\cdots+k$ means, add $k+1$ and the result will be $1+2+3+\cdots+(k+1)$. Think about how this could be done; for example, how should one define 973685! rigorously, i.e., without using $\cdots$?

Find a formula for $1+2+3+\cdots+k$.

M. Beck and R. Geoghegan, *The Art of Proof: Basic Training for Deeper Mathematics*, Undergraduate Texts in Mathematics, DOI 10.1007/978-1-4419-7023-7_4,

4.1 Examples

Assume that for each $j \in \mathbf{N}$, we are given some $x_j \in \mathbf{Z}$. We call the list of all the x_j a **sequence** (of integers) and we denote it by $(x_j)_{j=1}^{\infty}$. For some sequences, the subscript j does not start at 1 but at some other integer m, in which case we write $(x_j)_{j=m}^{\infty}$.

In Section 5.3 we will explain that a sequence is a function with domain **N**.

Example 4.1. Let $x_j := j^3 + j$. Then $x_1 = 2$, $x_2 = 10$, $x_3 = 30$, etc. The number x_k is called the kth **term** of the sequence.

Recall that when m divides n, there exists an integer a such that $n = ma$: we denote this integer a by $\frac{n}{m}$. In our next example we define a sequence in a way that should remind you of the principle of induction:

Note that $\frac{n}{m}$ is an integer here; this definition does not describe rational numbers (i.e., fractions). For example, our number system does not yet include $\frac{1}{2}$.

Example 4.2. ("$3x+1$ problem") Pick your favorite natural number m, and define the following sequence:

(i) Define $x_1 := m$, that is, set x_1 to be your favorite number.

(ii) Assuming x_n defined, define $x_{n+1} := \begin{cases} \frac{x_n}{2} & \text{if } x_n \text{ is even,} \\ 3x_n + 1 & \text{otherwise.} \end{cases}$

For example, if your favorite natural number is $m = 1$ then the sequence $(x_k)_{k=1}^{\infty}$ starts with $1, 4, 2, 1, 4, 2, 1, 4, 2, \ldots$. If your favorite number is $m = 3$, the sequence starts with $3, 10, 5, 16, 8, 4, 2, 1, 4, 2, \ldots$. It is a famous open conjecture that, no matter what $m \in \mathbf{N}$ you choose as the starting point, the sequence eventually takes on the value 1 (from which point the remainder of the sequence looks like $1, 4, 2, 1, 4, 2, 1, 4, 2, \ldots$).

Project 4.3. ("$x+1$ problem") We revise the $3x+1$ problem as follows: Pick your favorite natural number m, and define the following sequence:

(i) Define $x_1 := m$.

(ii) Assuming x_n defined, define $x_{n+1} := \begin{cases} \frac{x_n}{2} & \text{if } x_n \text{ is even,} \\ x_n + 1 & \text{otherwise.} \end{cases}$

Does this sequence eventually take on the value 1, no matter what $m \in \mathbf{N}$ one chooses as the starting point? Try to prove your assertion.

In Example 4.1 we have defined our sequence by a formula. In Example 4.2 and Project 4.3, the sequences are defined *recursively*. In a similar way we will now define sums, products, and factorials recursively:

Sum. Let $(x_j)_{j=1}^{\infty}$ be a sequence of integers. For each $k \in \mathbf{N}$, we want to define an integer called $\sum_{j=1}^{k} x_j$:

We sometimes write $x_1 + x_2 + \cdots + x_k$ when we mean $\sum_{j=1}^{k} x_j$.

(i) Define $\sum_{j=1}^{1} x_j$ to be x_1.

(ii) Assuming $\sum_{j=1}^{n} x_j$ already defined, we define $\sum_{j=1}^{n+1} x_j$ to be $\left(\sum_{j=1}^{n} x_j\right) + x_{n+1}$.

Product. Similarly, we define an integer called $\prod_{j=1}^{k} x_j$:

One can also write $x_1 x_2 \cdots x_k$ for $\prod_{j=1}^{k} x_j$.

(i) Define $\prod_{j=1}^{1} x_j := x_1$.

(ii) Assuming $\prod_{j=1}^{n} x_j$ defined, we define $\prod_{j=1}^{n+1} x_j := \left(\prod_{j=1}^{n} x_j\right) \cdot x_{n+1}$.

We denote the nonnegative integers by

$$\mathbf{Z}_{\geq 0} := \{m \in \mathbf{Z} : m \geq 0\}.$$

Factorial. As a third example, we define $k!$ ("k factorial") for all integers $k \geq 0$ by:

(i) Define $0! := 1$.

(ii) Assuming $n!$ defined (where $n \in \mathbf{Z}_{\geq 0}$), define $(n+1)! := (n!) \cdot (n+1)$.

In these examples, a new sequence is being defined step by step: the $(n+1)^{\text{th}}$ term can be written down only when you already know the nth term, so it may take substantial calculation to actually write down the $1{,}000{,}000^{\text{th}}$ term. Sometimes the rule that assigns a value y_j to each j is given by a formula, for example, $y_j = j^2 + 3$. Then you can see at a glance what answer the rule gives for any choice of j. But when, as in the above examples, the rule is given recursively, one could ask whether such a rule truly defines a sequence. The answer is *yes*. In fact, the legitimacy of this method of defining sequences can be deduced from our axioms, i.e., it is a theorem:

"at a glance" is an overstatement here: the number y_j might be so huge that it would take the biggest computer in the world to write it down, and even that computer might not be able to handle bigger y_j's; however, in a mathematical sense our sentence is correct.

Theorem 4.4. *A legitimate method of describing a sequence $(y_j)_{j=m}^{\infty}$ is:*

(i) *to name y_m and*

(ii) *to state a formula describing y_{n+1} in terms of y_n, for each $n \geq m$.*

Such a definition is called a *recursive definition*. In the above examples, Theorem 4.4 is being used in the special cases $m = 1$ ($3x+1$ problem, sum, product) and $m = 0$ (factorial).

Proof. The hard part here is to figure out what is to be proved. We are saying that if a sequence is described by (i) and (ii) then in principle any member of the sequence can be known. Let $P(k)$ be the statement "y_k can be known." We will use induction on k (Theorem 2.25).

Formally, we are claiming that the sequence, when regarded as a function, is well defined—see Section 5.4 for more on this. But here we wish to be less formal.

For the base case ($k = m$) we know that $P(m)$ is true because of (i).

For the induction step, assume that $P(n)$ is true, i.e., y_n can be known, for a particular n. Then (ii) allows us to deduce that y_{n+1} can also be known, and that finishes our induction. □

Proposition 4.5. *For all* $k \in \mathbf{Z}_{\geq 0}$, $k! \in \mathbf{N}$.

Power. Let b be a fixed integer. We define b^k for all integers $k \geq 0$ by:

(i) $b^0 := 1$.

(ii) Assuming b^n defined, let $b^{n+1} := b^n \cdot b$.

Proposition 4.6. *Let* $b \in \mathbf{Z}$ *and* $k, m \in \mathbf{Z}_{\geq 0}$.

(i) *If* $b \in \mathbf{N}$ *then* $b^k \in \mathbf{N}$.

(ii) $b^m b^k = b^{m+k}$.

(iii) $(b^m)^k = b^{mk}$.

Proof of (i) *and* (ii). We prove (i) by induction on $k \geq 0$; let $P(k)$ be the statement "$b^k \in \mathbf{N}$." The base case $P(0)$ follows with $b^0 = 1 \in \mathbf{N}$, by Proposition 2.3.

For the induction step, assume $b^n \in \mathbf{N}$ for some $n \in \mathbf{N}$; our goal is to conclude that $b^{n+1} \in \mathbf{N}$ also. By definition, $b^{n+1} = b^n \cdot b$; and since we know that both b and b^n are in $\mathbf{N}$, we conclude by Axiom 2.1(ii) that their product is also in $\mathbf{N}$.

Although the statement $b^m b^k = b^{m+k}$ is symmetric in m and k, we are doing induction on k for each fixed value of m.

For part (ii), fix $b \in \mathbf{Z}$ and $m \in \mathbf{Z}_{\geq 0}$. We will prove the statement $P(k): b^m b^k = b^{m+k}$ by induction on $k \geq 0$.

The base case $P(0)$ follows with $b^0 = 1$ and $m + 0 = m$, and so $P(0)$ simply states that

$$b^m \cdot 1 = b^m,$$

which holds by Axiom 1.3.

For the induction step, assume we know that $b^m b^n = b^{m+n}$ for some n; our goal is to conclude that $b^m b^{n+1} = b^{m+n+1}$. The left-hand side of this equation is, by definition,

$$b^m b^{n+1} = b^m \cdot b^n \cdot b\,, \tag{4.1}$$

whereas the right-hand side is, again by definition,

$$b^{m+n+1} = b^{m+n} \cdot b\,. \tag{4.2}$$

That the right-hand sides of (4.1) and (4.2) are equal follows with our induction assumption. □

The recursive definition of powers allows us to make more divisibility statements; here are a few examples.

Proposition 4.7. *For all $k \in \mathbf{N}$:*

(i) $5^{2k} - 1$ *is divisible by* 24;

(ii) $2^{2k+1} + 1$ *is divisible by* 3;

(iii) $10^k + 3 \cdot 4^{k+2} + 5$ *is divisible by* 9.

Proposition 4.8. *For all $k \in \mathbf{N}$, $4^k > k$.*

Project 4.9. Determine for which natural numbers $k^2 < 2^k$ and prove your answer.

Project 4.10. Come up with a recursive definition of the sum $\sum_{j=m}^{n} x_j$ for two integers $m \leq n$.

4.2 Finite Series

In this section, we will discuss some properties of sums (which we have defined recursively at the beginning of this chapter). We start with a few examples.

Proposition 4.11. *Let $k \in \mathbf{N}$.*

(i) $$\sum_{j=1}^{k} j = \frac{k(k+1)}{2}.$$

(ii) $$\sum_{j=1}^{k} j^2 = \frac{k(k+1)(2k+1)}{6}.$$

Proposition 4.11 implies that $k(k+1)$ is even and $k(k+1)(2k+1)$ is divisible by 6, for all $k \in \mathbf{N}$.

Project 4.12. Find (and prove) a formula for $\sum_{j=1}^{k} j^3$.

Setting $x_k := \sum_{j=1}^{k} j$ illustrates the fact that we can think of a sum like $\sum_{j=1}^{k} j$ as a sequence defined by a formula that varies with k. Such sequences, defined as sums, are called **finite series**. Here is another example, a **finite geometric series**:

We will discuss infinite series in Chapter 12.

Proposition 4.13. *For $x \neq 1$ and $k \in \mathbf{Z}_{\geq 0}$,* $$\sum_{j=0}^{k} x^j = \frac{1 - x^{k+1}}{1 - x}.$$

Project 4.14. With calculus, Proposition 4.13 can be used to generate formulas for $\sum_{j=0}^{k} j^m$ for $m = 1, 2, 3, 4, \ldots$, similar to those found in Proposition 4.11 and Project 4.12. Think about how this could be done.

Hint: start by differentiating both sides of Proposition 4.13 with respect to x and remember L'Hôpital's rule.

Proposition 4.15. (i) *Let $m \in \mathbf{Z}$ and let $(x_j)_{j=1}^{\infty}$ be a sequence in* **Z**. *Then for all $k \in \mathbf{N}$,*

$$m \cdot \left(\sum_{j=1}^{k} x_j \right) = \sum_{j=1}^{k} (m x_j) .$$

(ii) *If $x_j = 1$ for all $j \in \mathbf{N}$ then for all $k \in \mathbf{N}$,*

$$\sum_{j=1}^{k} x_j = k .$$

(iii) *If $x_j = n \in \mathbf{Z}$ for all $j \in \mathbf{N}$ then for all $k \in \mathbf{N}$,*

$$\sum_{j=1}^{k} x_j = kn .$$

Proposition 4.16. *Let $(x_j)_{j=1}^{\infty}$ and $(y_j)_{j=1}^{\infty}$ be sequences in* **Z**, *and let $a,b,c \in \mathbf{Z}$ be such that $a \leq b < c$.*

(i) $\displaystyle\sum_{j=a}^{c} x_j = \sum_{j=a}^{b} x_j + \sum_{j=b+1}^{c} x_j .$

(ii) $\displaystyle\sum_{j=a}^{b} (x_j + y_j) = \left(\sum_{j=a}^{b} x_j \right) + \left(\sum_{j=a}^{b} y_j \right).$

Can you see that both expressions mean $x_a + \cdots + x_b$? This way of reorganizing a sum can be useful. See, for example, the proof of Theorem 4.21.

Proposition 4.17. *Let $(x_j)_{j=1}^{\infty}$ be a sequence in* **Z**, *and let $a,b,r \in \mathbf{Z}$ be such that $a \leq b$. Then*

$$\sum_{j=a}^{b} x_j = \sum_{j=a+r}^{b+r} x_{j-r} .$$

Proposition 4.18. *Let $(x_j)_{j=1}^{\infty}$ and $(y_j)_{j=1}^{\infty}$ be sequences in* **Z** *such that $x_j \leq y_j$ for all $j \in \mathbf{N}$. Then for all $k \in \mathbf{N}$,*

$$\sum_{j=1}^{k} x_j \leq \sum_{j=1}^{k} y_j .$$

4.3 Fishing in a Finite Pool

We started this chapter with an infinite sequence $(x_j)_{j=m}^{\infty}$, and for each $k \geq m$ we defined numbers such as $\sum_{j=m}^{k} x_j$ and $\prod_{j=m}^{k} x_j$. Sometimes you start with a *finite* sequence $(x_j)_{j=m}^{M}$, i.e., a list of numbers

$$x_m, x_{m+1}, x_{m+2}, \ldots, x_{M-1}, x_M .$$

Convention: if $M = m$, this list is x_m; if $M = m+1$, this list is x_m, x_{m+1}; and so on.

Then our definitions of $\sum_{j=m}^{k} x_j$ and $\prod_{j=m}^{k} x_j$ make sense for all $m \le k \le M$. In words: your initial pool of numbers can be finite.

This remark also applies to induction. If some statement $P(k)$ makes sense only for $m \le k \le M$, you can prove $P(k)$ for $m \le k \le M$ by proving $P(m)$, and $P(n) \Rightarrow P(n+1)$ when $m \le n < M$.

4.4 The Binomial Theorem

Theorem 4.19. *Let $k, m \in \mathbf{Z}_{\ge 0}$, where $m \le k$. Then $m!(k-m)!$ divides $k!$.*

Thus $\frac{k!}{m!(k-m)!}$ is an integer; it is customary to denote this integer by the symbol $\binom{k}{m}$ and call it a **binomial coefficient**.

Proof. For each $k \in \mathbf{Z}_{\ge 0}$ we let $P(k)$ be the statement

$$\text{for all } 0 \le m \le k, \text{ there exists } j \in \mathbf{Z} \text{ such that } k! = j\, m!(k-m)! .$$

Theorem 4.19 says that $P(k)$ is true for all $k \in \mathbf{Z}_{\ge 0}$, so that is what we will prove by induction on k. That is, we will prove that $P(0)$ is true and $P(n)$ implies $P(n+1)$.

Before doing this, we note that for any given k, the statement $P(k)$ contains $k+1$ pieces of information, one for each value of m; in particular, the number j mentioned in our statement of $P(k)$ depends on m. So, for example, $P(2)$ says that

- there exists $j_0 \in \mathbf{Z}$ such that $2! = j_0\, 0!\,2!$
- there exists $j_1 \in \mathbf{Z}$ such that $2! = j_1\, 1!\,1!$
- there exists $j_2 \in \mathbf{Z}$ such that $2! = j_2\, 2!\,0!$.

What are j_0, j_1, j_2?

This statement $P(2)$ is apparently true, but $P(1{,}000{,}000)$ is not so obvious.

Now we give the induction proof. The statement $P(0)$ is true: it says that there exists an integer j such that $0! = j\,0!\,0!$.

In proving that $P(n)$ implies $P(n+1)$, we first note that the extreme cases $m = 0$ and $m = n+1$ of $P(n+1)$ are simple; for both of these the required j is 1:

$$(n+1)! = 1 \cdot 0!\,(n+1)! \qquad \text{and} \qquad (n+1)! = 1 \cdot (n+1)!\,0! .$$

So we are to prove the remaining cases of $P(n+1)$, that is,

$$\text{for all } 1 \le m \le n, \text{ there exists } j \in \mathbf{Z} \text{ such that } (n+1)! = j\, m!(n+1-m)! .$$

For a particular m we will use the m and $m-1$ cases of the (assumed true) statement $P(n)$; i.e., there exist integers a and b such that

Can you see why we separated the cases $m = 0$ and $m = n+1$?

$$n! = a\,(m-1)!\,(n-m+1)! \qquad \text{and} \qquad n! = b\,m!\,(n-m)!\,.$$

But then

$$\begin{aligned}(n+1)! &= n!\,(m+n+1-m)\\ &= n!\,m + n!\,(n+1-m)\\ &= a\,(m-1)!\,(n-m+1)!\,m + b\,m!\,(n-m)!\,(n+1-m)\\ &= (a+b)\,m!\,(n-m+1)!\end{aligned}$$

which completes our induction step: the number $a+b$ is the required j. $\square$

In our proof of Theorem 4.19—more precisely, in the second-to-last line—we deduced the following recursive identity for the binomial coefficients:

Corollary 4.20. *For* $1 \le m \le k$, $\binom{k+1}{m} = \binom{k}{m-1} + \binom{k}{m}$.

You have most certainly seen this recursion in disguise, namely, when you discussed binomial expansions in school:

$$\begin{aligned}(a+b)^2 &= a^2 + 2ab + b^2,\\ (a+b)^3 &= a^3 + 3a^2b + 3ab^2 + b^3, \ \text{etc.}\end{aligned}$$

Your teacher may have explained that one can obtain the coefficients from **Pascal's triangle**:

$\binom{k}{m}$ is the mth term of the kth row, where k and m start at 0.

$$\begin{array}{ccccccccccccc} &&&&&&1&&&&&& \\ &&&&&1&&1&&&&& \\ &&&&1&&2&&1&&&& \\ &&&1&&3&&3&&1&&& \\ &&1&&4&&6&&4&&1&& \\ &1&&5&&10&&10&&5&&1& \\ 1&&6&&15&&20&&15&&6&&1 \\ &&&&&&\vdots&&&&&& \end{array}$$

Corollary 4.20 tells us that each entry in a row is the sum of its two neighbors in the previous row. Here is the general form of your high-school theorem.

Theorem 4.21 (Binomial theorem for integers). *If* $a, b \in \mathbf{Z}$ *and* $k \in \mathbf{Z}_{\ge 0}$ *then*

$$(a+b)^k = \sum_{m=0}^{k} \binom{k}{m} a^m\, b^{k-m}.$$

Proof. We prove this by induction on $k \geq 0$. The base case follows with $x^0 = 1$ (which we use for $x = a+b$, a, and b) and $\binom{0}{0} = 1$.

For the induction step, assume that $(a+b)^n = \sum_{m=0}^{n} \binom{n}{m} a^m b^{n-m}$ for some $n \geq 0$. Then

$$\sum_{m=0}^{n+1} \binom{n+1}{m} a^m b^{n+1-m}$$

$$= \binom{n+1}{0} a^0 b^{n+1} + \sum_{m=1}^{n} \binom{n+1}{m} a^m b^{n+1-m} + \binom{n+1}{n+1} a^{n+1} b^0$$

(by Proposition 4.16(i))

$$= b^{n+1} + \sum_{m=1}^{n} \left(\binom{n}{m-1} + \binom{n}{m} \right) a^m b^{n+1-m} + a^{n+1}$$

(by definition of powers/binomial coefficients and Corollary 4.20)

$$= b^{n+1} + \sum_{m=1}^{n} \binom{n}{m-1} a^m b^{n+1-m} + \sum_{m=1}^{n} \binom{n}{m} a^m b^{n+1-m} + a^{n+1}$$

(by distributivity and Proposition 4.16(ii))

$$= b^{n+1} + \sum_{m=0}^{n-1} \binom{n}{m} a^{m+1} b^{n+1-(m+1)} + \sum_{m=1}^{n} \binom{n}{m} a^m b^{n+1-m} + a^{n+1}$$

We are using Proposition 4.17 with $r = 1$.

(by Proposition 4.17 applied to the first sum)

$$= \sum_{m=0}^{n} \binom{n}{m} a^{m+1} b^{n+1-(m+1)} + \sum_{m=0}^{n} \binom{n}{m} a^m b^{n+1-m}$$

(by combining a^{n+1} with the first sum and b^{n+1} with the second sum, using Proposition 4.16(i))

$$= a \sum_{m=0}^{n} \binom{n}{m} a^m b^{n-m} + b \sum_{m=0}^{n} \binom{n}{m} a^m b^{n-m}$$

(by definition of powers and Proposition 4.15(i))

$$= a(a+b)^n + b(a+b)^n$$

(by the induction hypothesis)

$$= (a+b)(a+b)^n$$

(by distributivity)

$$= (a+b)^{n+1}$$

(by definition of powers). □

An immediate corollary of the binomial theorem, namely the special case $a = b = 1$, gives another relation among the binomial coefficients:

Corollary 4.22. *For* $k \in \mathbf{Z}_{\geq 0}$, $\sum_{m=0}^{k} \binom{k}{m} = 2^k$.

A slight variation of the binomial theorem is the general product formula of calculus—although this looks like a completely different topic at first sight.

Project 4.23 (Leibniz's formula). Consider an operation denoted by $'$ that is applied to symbols such as u, v, w. Assume that the operation $'$ satisfies the following axioms:

$$\begin{aligned} (u+v)' &= u' + v', \\ (uv)' &= uv' + u'v, \\ (cu)' &= cu', \quad \text{where } c \text{ is a constant.} \end{aligned} \tag{4.3}$$

Define $w^{(k)}$ recursively by

(i) $w^{(0)} := w$.

(ii) Assuming $w^{(n)}$ defined (where $n \in \mathbf{Z}_{\geq 0}$), define $w^{(n+1)} := (w^{(n)})'$.

This formula was found by Gottfried Leibniz (1646–1716)—a codiscoverer (with Isaac Newton) of calculus—in 1678.

Prove:

$$(uv)^{(k)} = \sum_{m=0}^{k} \binom{k}{m} u^{(m)} v^{(k-m)}.$$

You know a case of this operation $'$, namely, u, v, and w are functions and u', v', and w' are the derivatives of these functions. Differentiation satisfies our axioms (4.3). In this context Leibniz's formula calculates the kth derivative of the product of two functions. It is interesting that we are using a mathematical method unrelated to calculus to prove a theorem in calculus.

4.5 A Second Form of Induction

Recall the principle of induction—first form (Theorem 2.17): Let $P(k)$ be a statement depending on a variable $k \in \mathbf{N}$. In order to prove the statement "$P(k)$ is true for all $k \in \mathbf{N}$" it is sufficient to prove:

(i) $P(1)$ is true and

(ii) if $P(n)$ is true then $P(n+1)$ is true.

There is a second form of this, a trick that people call **strong induction**. It is just another way of stating the same idea, but it sounds different, and it can be useful:

Theorem 4.24 (Principle of mathematical induction—second form). *Let $P(k)$ be a statement depending on a variable $k \in \mathbf{N}$. In order to prove the statement "$P(k)$ is true for all $k \in \mathbf{N}$" it is sufficient to prove:*

(i) *$P(1)$ is true;*

(ii) *if $P(j)$ is true for all integers j such that $1 \le j \le n$, then $P(n+1)$ is true.*

One way to prove this theorem is by the first form of induction (Theorem 2.17)—try proving the statement "$P(j)$ is true for all integers j such that $1 \le j \le k$" by induction on k.

Project 4.25 (Principle of mathematical induction—second form revisited). State and prove the analogue of Theorem 2.25 for this second form of induction.

Project 4.26. A sequence $(x_j)_{j=0}^{\infty}$ satisfies

$$x_1 = 1 \qquad \text{and for all } m \ge n \ge 0, \qquad x_{m+n} + x_{m-n} = \tfrac{1}{2}\left(x_{2m} + x_{2n}\right).$$

Find a formula for x_j. Prove that your formula is correct.

4.6 More Recursions

The strong induction principle allows us to define a new kind of recursive sequence. Here is a famous example:

Example 4.27. The **Fibonacci numbers** $(f_j)_{j=1}^{\infty}$ are defined by $f_1 := 1$, $f_2 := 1$, and

$$f_n = f_{n-1} + f_{n-2} \qquad \text{for } n \ge 3. \tag{4.4}$$

The Fibonacci numbers are named after Leonardo of Pisa (c. 1170–c. 1250), also known as Leonardo Fibonacci.

A **recurrence relation** for a sequence is a formula that describes the nth term of the sequence using terms with smaller subscripts. Equation (4.4) for the Fibonacci numbers is an example.

Project 4.28. Calculate f_{13}.

The Fibonacci numbers obey a surprising formula—assuming for a moment that we know about real numbers such as $\sqrt{5}$.

Note that Proposition 4.29 implies that the strange expression on the right-hand side is always an integer.

Proposition 4.29. *The kth Fibonacci number is given directly by the formula*

$$f_k = \frac{1}{\sqrt{5}}\left(\left(\frac{1+\sqrt{5}}{2}\right)^k - \left(\frac{1-\sqrt{5}}{2}\right)^k\right).$$

Proof. Let $a = \frac{1+\sqrt{5}}{2}$ and $b = \frac{1-\sqrt{5}}{2}$. We prove $P(k) : f_k = \frac{1}{\sqrt{5}}\left(a^k - b^k\right)$ by (strong) induction on $k \in \mathbf{N}$. First, we check $P(1)$ and $P(2)$, for which the formula gives $f_1 = 1 = f_2$.

We need to check two base cases, because the recursion formula for f_k involves the two previous sequence members.

For the induction step, assume that $P(j)$ is true for $1 \leq j \leq n$, for some $n \geq 2$. Then, by definition of the Fibonacci sequence and the induction assumption,

$$\begin{aligned} f_{n+1} &= f_n + f_{n-1} \\ &= \frac{1}{\sqrt{5}}\left(a^n - b^n\right) + \frac{1}{\sqrt{5}}\left(a^{n-1} - b^{n-1}\right) \\ &= \frac{1}{\sqrt{5}}\left(a^n + a^{n-1} - b^n - b^{n-1}\right) \\ &= \frac{1}{\sqrt{5}}\left(a^{n-1}(a+1) - b^{n-1}(b+1)\right) \\ &= \frac{1}{\sqrt{5}}\left(a^{n-1}a^2 - b^{n-1}b^2\right) \\ &= \frac{1}{\sqrt{5}}\left(a^{n+1} - b^{n+1}\right). \end{aligned}$$

Here we used

$$a+1 = \frac{3+\sqrt{5}}{2} = \frac{1+2\sqrt{5}+5}{4} = a^2$$

and

$$b+1 = \frac{3-\sqrt{5}}{2} = \frac{1-2\sqrt{5}+5}{4} = b^2. \qquad \square$$

The Fibonacci numbers have numerous interesting properties; to give a flavor we present a few here (try to prove them without using Proposition 4.29):

Proposition 4.30. *For all $k, m \in \mathbf{N}$, where $m \geq 2$,*

$$f_{m+k} = f_{m-1} f_k + f_m f_{k+1}.$$

Proposition 4.31. *For all $k \in \mathbf{N}$, $f_{2k+1} = f_k^2 + f_{k+1}^2$.*

Proposition 4.32. *For all $k, m \in \mathbf{N}$, f_{mk} is divisible by f_m.*

Project 4.33. How many ways are there to order the numbers $1, 2, \ldots, 20$ in a row so that the first number is 1, the last number is 20, and each pair of consecutive numbers differ by at most 2?

Review Question. Do you understand what it means to define something recursively?

Weekly reminder: Reading mathematics is not like reading novels or history. You need to think slowly about every sentence. Usually, you will need to reread the same material later, often more than one rereading.

This is a short book. Its core material occupies about 140 pages. Yet it takes a semester for most students to master this material. In summary: read line by line, not page by page.

Chapter 5
Underlying Notions in Set Theory

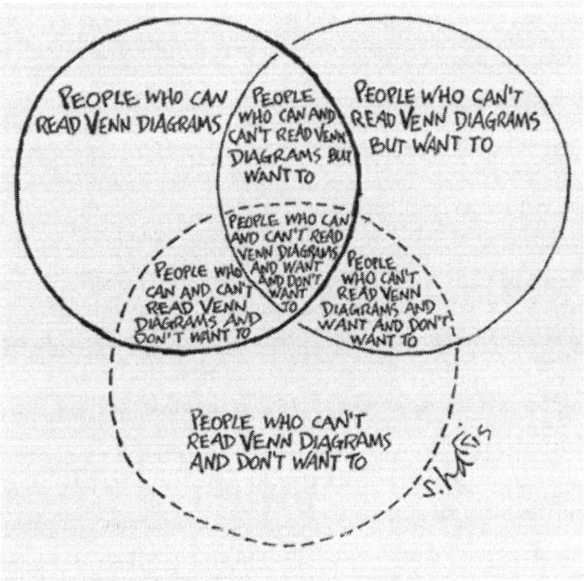

Before You Get Started. Come up with real-life examples of sets that behave like the ones in the above picture; for example, you might think of friends of yours that could be grouped according to certain characteristics—those younger than 20, those who are female, etc. Carefully label your picture. Make your example rich enough that all of the regions in the picture have members.

M. Beck and R. Geoghegan, *The Art of Proof: Basic Training for Deeper Mathematics*, Undergraduate Texts in Mathematics, DOI 10.1007/978-1-4419-7023-7_5,

5.1 Subsets and Set Equality

Pay attention to the difference between an element of a set S and a subset of a set S. A subset of S is a set, while an element of S is one of the things that is in the set S.

A **set** is a collection of "things" usually called **elements** or **members**. The notation $x \in A$ means that x is a member of (an element of) the set A. The negation of $x \in A$ is written $x \notin A$. It means that x is not a member of A.

As we saw in Chapter 2, we write $A \subseteq B$ (A is a subset of B) when every member of A is a member of B, i.e.,

$$x \in A \Rightarrow x \in B.$$

The symbol $\supseteq$ is also used when we want to read from right to left: $B \supseteq A$ means $A \subseteq B$.

Proposition 5.1. *Let A, B, C be sets.*

(i) $A \subseteq A$.

(ii) *If $A \subseteq B$ and $B \subseteq C$ then $A \subseteq C$.*

Proof. (i) $A \subseteq A$ means "if $x \in A$ then $x \in A$," which is a true statement.

(ii) Assume $A \subseteq B$ and $B \subseteq C$. We need to show that if $x \in A$ then $x \in C$. Given $x \in A$, $A \subseteq B$ implies that $x \in B$. Since $B \subseteq C$, this implies that $x \in C$. □

Another concept, already introduced in Chapter 2, is set equality: We write $A = B$ when A and B are the same set, i.e., when A and B have precisely the same members, i.e., when

$$A \subseteq B \qquad \text{and} \qquad B \subseteq A. \tag{5.1}$$

Note that equality of sets has a different flavor from equality of numbers. To prove that two sets are equal often involves hard work—we have to establish the two subset relations in (5.1).

Sometimes the same set can be described in two apparently different ways. For example, let A be the set of all integers of the form $7m+1$, where $m \in \mathbf{Z}$, and let B be the set of all integers of the form $7n-6$, where $n \in \mathbf{Z}$. We write this as

$$A = \{7m+1 : m \in \mathbf{Z}\} \qquad \text{and} \qquad B = \{7n-6 : n \in \mathbf{Z}\}.$$

This proposition might be too simple to be interesting. We have included it to illustrate how one proves that two sets are equal.

Proposition 5.2. $\{7m+1 : m \in \mathbf{Z}\} = \{7n-6 : n \in \mathbf{Z}\}$.

Proof. We must prove that $A \subseteq B$ and $B \subseteq A$.

The first statement means $x \in A \Rightarrow x \in B$. So let $x \in A$. Then, for some $m \in \mathbf{Z}$, $x = 7m+1$. But $7m+1 = 7(m+1)-6$, and so we can set $n = m+1$, which gives $x = 7n-6$; thus $x \in B$. This proves $A \subseteq B$.

Conversely, let $x \in B$. Then, for some $n \in \mathbf{Z}$, $x = 7n - 6$. But $7n - 6 = 7(n-1) + 1$; setting $m = n - 1$ gives $x = 7m + 1$, and so $x \in A$. This proves $B \subseteq A$ and establishes our desired set equality. □

Template for proving $A = B$. Prove that $A \subseteq B$ and $B \subseteq A$.

Project 5.3. Define the following sets:

$$\begin{aligned} A &:= \{3x : x \in \mathbf{N}\}, \\ B &:= \{3x + 21 : x \in \mathbf{N}\}, \\ C &:= \{x + 7 : x \in \mathbf{N}\}, \\ D &:= \{3x : x \in \mathbf{N} \text{ and } x > 7\}, \\ E &:= \{x : x \in \mathbf{N}\}, \\ F &:= \{3x - 21 : x \in \mathbf{N}\}, \\ G &:= \{x : x \in \mathbf{N} \text{ and } x > 7\}. \end{aligned}$$

Determine which of the following set equalities are true. If a statement is true, prove it. If it is false, explain why this set equality does not hold.

(i) $D = E$.

(ii) $C = G$.

(iii) $D = B$.

The sets A and F will make an appearance in Project 5.11.

Here are some facts about equality of sets:

Proposition 5.4. *Let A, B, C be sets.*

(i) *$A = A$.*

(ii) *If $A = B$ then $B = A$.*

(iii) *If $A = B$ and $B = C$ then $A = C$.*

These three properties should look familiar—we mentioned them already in Section 1.1 when we talked about equality of two integers. We called the properties *reflexivity*, *symmetry*, and *transitivity*, respectively.

We will see these properties again in Section 6.1.

Project 5.5. When reading or writing a set definition, pay attention to what is a variable inside the set definition and what is not a variable. As examples, how do the following pairs of sets differ?

(i) $S := \{m : m \in \mathbf{N}\}$ and $T_m := \{m\}$ for a specified $m \in \mathbf{N}$.

(ii) $U := \{my : y \in \mathbf{Z}, m \in \mathbf{N}, my > 0\}$ and $V_m := \{my : y \in \mathbf{Z}, my > 0\}$ for a specified $m \in \mathbf{N}$.

The subscripts on T_m, V_m, W_m are not necessary, but this notation is often useful to emphasize the fact that m is a constant.

(iii) V_m and $W_m := \{my : y \in \mathbf{Z}, y > 0\}$ for a specified $m \in \mathbf{Z}$.

Find the simplest possible way of writing each of these sets.

The **empty set**, denoted by $\varnothing$, has the feature that $x \in \varnothing$ is never true. We allow ourselves to say *the* empty set because there is only one set with this property:

Proposition 5.6 asserts the uniqueness of $\varnothing$. The existence of $\varnothing$ is one of the hidden assumptions mentioned in Section 1.4.

Proposition 5.6. *If the sets $\varnothing_1$ and $\varnothing_2$ have the property that $x \in \varnothing_1$ is never true and $x \in \varnothing_2$ is never true, then $\varnothing_1 = \varnothing_2$.*

Proof. Assume that the sets $\varnothing_1$ and $\varnothing_2$ have the property that $x \in \varnothing_1$ is never true and $x \in \varnothing_2$ is never true. Suppose (by means of contradiction) that $\varnothing_1 \neq \varnothing_2$, that is, either $\varnothing_1 \not\subseteq \varnothing_2$ or $\varnothing_1 \not\supseteq \varnothing_2$. We first consider the case $\varnothing_1 \not\subseteq \varnothing_2$. This means there is some $x \in \varnothing_1$ such that $x \notin \varnothing_2$. But that cannot be, since there is no $x \in \varnothing_1$. The other case, $\varnothing_1 \not\supseteq \varnothing_2$, is dealt with similarly. □

Proposition 5.7. *The empty set is a subset of every set, that is, for every set S, $\varnothing \subseteq S$.*

Project 5.8. Read through the proof of Proposition 5.1 having in mind that A is empty. Then there exists no x that is in A. Do you see why the proof still holds?

5.2 Intersections and Unions

The **intersection** of two sets A and B is

$$A \cap B = \{x : x \in A \text{ and } x \in B\}.$$

The **union** of A and B is

$$A \cup B = \{x : x \in A \text{ or } x \in B\}.$$

The set operations $\cap$ and $\cup$ give us alternative ways of writing certain sets. Here are two examples:

When two sets A and B satisfy $A \cap B = \varnothing$, we say that A and B are disjoint.

Example 5.9. $\{3x+1 : x \in \mathbf{Z}\} \cap \{3x+2 : x \in \mathbf{Z}\} = \varnothing$.

Example 5.10.
$\{2x : x \in \mathbf{Z}, 3 \leq x\} = \{x \in \mathbf{Z} : 5 \leq x\} \cap \{x \in \mathbf{Z} : x \text{ is even}\}$.

Project 5.11. This is a continuation of Project 5.3, and so the following names refer to the sets defined in Project 5.3. Again, determine which of the following set equalities are true. If a statement is true, prove it. If it is false, explain why this set equality does not hold.

(i) $A \cap E = B$.

(ii) $A \cap C = B$.

(iii) $E \cap F = A$.

Project 5.12. Determine which of the following statements are true for all sets A, B, and C. If a double implication fails, determine whether one or the other of the possible implications holds. If a statement is true, prove it. If it is false, provide a counterexample.

"Providing a counterexample" here means coming up with a specific example of a set triple A,B,C that violates the statement.

(i) $C \subseteq A$ and $C \subseteq B \iff C \subseteq (A \cup B)$.

(ii) $C \subseteq A$ or $C \subseteq B \iff C \subseteq (A \cup B)$.

(iii) $C \subseteq A$ and $C \subseteq B \iff C \subseteq (A \cap B)$.

(iv) $C \subseteq A$ or $C \subseteq B \iff C \subseteq (A \cap B)$.

For two sets A and B, we define the **set difference**

$$A - B = \{x : x \in A \text{ and } x \notin B\}.$$

Another commonly used notation for set difference is $A \setminus B$.

Given a set $A \subseteq X$, we define the **complement** of A in X to be $X - A$. If the bigger set X is clear from the context, one often writes A^c for the complement of A (in X).

Example 5.13. Recall that the even integers are those integers that are divisible by 2. The **odd** integers are defined to be those integers that are not even. Thus the set of odd integers is the complement of the set of even integers.

Proposition 5.14. *Let $A, B \subseteq X$.*

$$A \subseteq B \qquad \textit{if and only if} \qquad B^c \subseteq A^c.$$

Theorem 5.15 (De Morgan's laws). *Given two subsets $A, B \subseteq X$,*

$$(A \cap B)^c = A^c \cup B^c \qquad \textit{and} \qquad (A \cup B)^c = A^c \cap B^c.$$

In words: the complement of the intersection is the union of the complements and the complement of the union is the intersection of the complements.

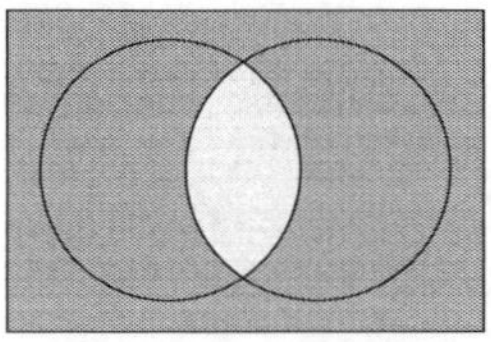

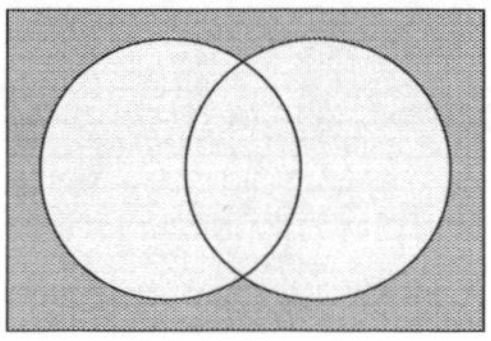

Here are two pictures of De Morgan's equalities.

Project 5.16. Someone tells you that the following equalities are true for all sets A, B, C. In each case, either prove the claim or provide a counterexample.

(i) $A - (B \cup C) = (A - B) \cup (A - C)$.

(ii) $A \cap (B - C) = (A \cap B) - (A \cap C)$.

As another example of a recursive construction, we invite you to explore unions and intersections of an arbitrary number of sets.

Project 5.17 (Unions and intersections). Given sets $A_1, A_2, A_3, \ldots$, develop recursive definitions for

$$\bigcup_{j=1}^{k} A_j \quad \text{and} \quad \bigcap_{j=1}^{k} A_j .$$

Find and prove an extension of De Morgan's laws (Theorem 5.15) for these unions and intersections.

Logical paradoxes arise when one treats the "set of all sets" as a set. The "set" R in Project 5.18 is an indication of the problem. These logical issues do not cause difficulties in the mathematics discussed in this book.

Proposition 5.7 says that the empty set $\varnothing$ is "extreme" in that it is the smallest possible set. Thus $S \neq \varnothing$ if and only if there exists an x such that $x \in S$. One would like to go to the other extreme and define a set that contains "everything"; however, there is no such set.

Project 5.18. Let $R = \{X : X \text{ is a set and } X \notin X\}$. Is the statement $R \in R$ true or false?

5.3 Cartesian Products

Let A and B be sets. From them we obtain a new set

$$A \times B := \{(a,b) : a \in A \text{ and } b \in B\} .$$

Cartesian products are named after René Descartes (1596–1650), who used this concept in his development of analytic geometry.

We call (a,b) an **ordered pair**. The set $A \times B$ is called the **(Cartesian) product** of A and B. It is the set of all ordered pairs whose first entry is a member of A and whose second entry is a member of B.

Example 5.19. $(3,-2)$ is an ordered pair of integers, and $\mathbf{Z} \times \mathbf{Z}$ denotes the set of all ordered pairs of integers. (Draw a picture.)

Notice that when $A \neq B$, $A \times B$ and $B \times A$ are different sets.

Proposition 5.20. *Let A, B, C be sets.*

(i) $A \times (B \cup C) = (A \times B) \cup (A \times C)$.

(ii) $A \times (B \cap C) = (A \times B) \cap (A \times C)$.

Project 5.21. Let A, B, C, D be sets. Decide whether each of the following statements is true or false; in each case prove the statement or give a counterexample.

(i) $(A \times B) \cup (C \times D) = (A \cup C) \times (B \cup D)$.

(ii) $(A \times B) \cap (C \times D) = (A \cap C) \times (B \cap D)$.

5.4 Functions

We come to one of the most important ideas in mathematics. There is an informal definition and a more abstract definition of the concept of a function. We give both.

First Definition. A **function** consists of

- a set A called the **domain** of the function;
- a set B called the **codomain** of the function;
- a rule f that assigns to each $a \in A$ an element $f(a) \in B$.

A useful shorthand for this is $f : A \to B$.

This notation suggests that the function f picks up each $a \in A$ and carries it over to B, placing it precisely on top of an element $f(a) \in B$.

Example 5.22. $f : \mathbf{Z} \to \mathbf{Z}$ given by $f(n) = n^3 + 1$.

Example 5.23. Every sequence $(x_j)_{j=1}^{\infty}$ is a function with domain $\mathbf{N}$, where we write x_j instead of $f(j)$.

The **graph** of $f : A \to B$ is

$$\Gamma(f) = \{(a,b) \in A \times B : b = f(a)\} .$$

Sometimes mathematicians ask whether a function is well defined. What they mean is this: "Does the rule you propose really assign to each element of the domain one and only one value in the codomain?"

Project 5.24. Discuss how much of this concept coincides with the notion of the graph of $f(x)$ in your calculus courses.

A possible objection to our first definition is that we used the undefined words *rule* and *assigns*. To avoid this, we offer the following alternative definition of a function through its graph:

Second Definition. A **function** with **domain** A and **codomain** B is a subset Γ of $A \times B$ such that for each $a \in A$ there is one and only one element of Γ whose first entry is a. If $(a,b) \in \Gamma$, we write $b = f(a)$.

Project 5.25. Discuss our two definitions of function. What are the advantages and disadvantages of each? Compare them with the definition you learned in calculus.

Example 5.26. A binary operation on a set A is a function $f : A \times A \to A$. For example, Axiom 1.1 could be restated as follows: There are two functions plus: $\mathbf{Z} \times \mathbf{Z} \to \mathbf{Z}$ and times: $\mathbf{Z} \times \mathbf{Z} \to \mathbf{Z}$ such that for all integers m, n, and p,

$$\begin{aligned}
\text{plus}(m,n) &= \text{plus}(n,m) \\
\text{plus}\left(\text{plus}(m,n),p\right) &= \text{plus}\left(m,\text{plus}(n,p)\right) \\
\text{times}\left(m,\text{plus}(n,p)\right) &= \text{plus}\left(\text{times}(m,n),\text{times}(m,p)\right) \\
\text{times}(m,n) &= \text{times}(n,m) \\
\text{times}\left(\text{times}(m,n),p\right) &= \text{times}\left(m,\text{times}(n,p)\right).
\end{aligned}$$

Review Question. Do you understand the difference between $\in$ and $\subseteq$?

Weekly reminder: Reading mathematics is not like reading novels or history. You need to think slowly about every sentence. Usually, you will need to reread the same material later, often more than one rereading.

This is a short book. Its core material occupies about 140 pages. Yet it takes a semester for most students to master this material. In summary: read line by line, not page by page.

Chapter 6
Equivalence Relations and Modular Arithmetic

Mathematics is the art of giving the same name to different things.
Jules Henri Poincaré (1854–1912)

In this chapter we discuss equivalence relations and illustrate how they apply to basic number theory. Equivalence relations are of fundamental importance in mathematics. The epigraph by Poincaré at the top of this page "says it all," but perhaps an explanation of what he had in mind would help. As an example, consider the set of all members of a club. Group together those whose birthdays occur in the same month. Two members are thus declared "equivalent" if they belong to the same group—if their birthdays are in the same month—and the set of people in one group is called an "equivalence class." Every club member belongs to one and only one equivalence class.

Before You Get Started. Consider our birthday groups. What properties do you notice about our birth-month equivalence relation and about the equivalence classes? Can you think of other examples in which we group things together? Does this lead to a guess as to how you might define equivalence relations in general?

M. Beck and R. Geoghegan, *The Art of Proof: Basic Training for Deeper Mathematics*,
Undergraduate Texts in Mathematics, DOI 10.1007/978-1-4419-7023-7_6,

6.1 Equivalence Relations

A **relation** on a set A is a subset of $A \times A$. Given a relation $R \subseteq A \times A$, we often write $x \sim y$ instead of $(x,y) \in R$ and we say that x **is related to** y (by the relation R).

Example 6.1. Some familiar examples of relations $a \sim b$ in $\mathbf{Z}$ are:

What we mean here is that any of $=, <, \leq$, and divides can play the role of $\sim$.

- $a = b$
- $a < b$
- $a \leq b$
- a divides b.

Example 6.2. The graph of a function $f : A \to A$ is a special case of a relation (for which there is exactly one $(x,y) \in R$ for each $x \in A$).

The relation $R \subseteq A \times A$ is an **equivalence relation** if it has the following three properties:

(i) $a \sim a$ for all $a \in A$. *(reflexivity)*

(ii) $a \sim b$ implies $b \sim a$. *(symmetry)*

(iii) $a \sim b$ and $b \sim c$ imply $a \sim c$. *(transitivity)*

Given an equivalence relation $\sim$ on A, the **equivalence class** of $a \in A$ is

$$[a] := \{b \in A : b \sim a\}.$$

Example 6.3. Of the relations $a \sim b$ in Example 6.1, only the one defined by $a = b$ is an equivalence relation.

If nobody's birthday occurs in February, then there is no February equivalence class: in other words, we do not count the empty set as an equivalence class. So there are at most 12 equivalence classes in this example, perhaps fewer than 12.

In our birthday example, Jasper $\sim$ Jennifer if and only if their birthdays occur in the same month. The equivalence class of Jennifer contains all the club members that share their birth month with Jennifer. One might as well label this equivalence class with the name of this month, but in mathematics one usually writes [Jennifer], i.e., the equivalence class is labeled by one of its members.

But suppose instead we declared two club members to be "equivalent" if their first names begin with the same letter: that would be a different equivalence relation on the same set.

Hint: reread carefully the definition of an equivalence relation.

It might happen (coincidences do happen) that in this particular club two members have birthdays in the same month if and only if their first names begin with the same letter. In that case do we have different equivalence relations or the same equivalence relation? The answer is, the same. Explain this answer to someone else.

Proposition 6.4. *Given an equivalence relation* $\sim$ *on a set* A *and* $a, b \in A$*,*

(i) $a \in [a]$.

(ii) $a \sim b$ *if and only if* $[a] = [b]$.

Proof. (i) By the reflexivity property of $\sim$ we have $a \sim a$, that is, $a \in [a]$.

(ii) Assume $a \sim b$. To show that $[a] = [b]$ we need to prove that $[a] \subseteq [b]$ and $[b] \subseteq [a]$.

If $c \in [a]$ then $c \sim a$, and so by transitivity of $\sim$ we have $c \sim b$, that is, $c \in [b]$. This proves $[a] \subseteq [b]$.

If $c \in [b]$ then $c \sim b$; now we use both symmetry and transitivity of $\sim$ to conclude that $c \sim a$, and so $c \in [a]$. This proves $[b] \subseteq [a]$.

Conversely, assume $[a] = [b]$. By part (i), $a \in [a] = [b]$, so $a \in [b]$. But this means $a \sim b$. □

This proposition implies that for every $a \in A$, there is a *unique* equivalence class (namely $[a]$) that contains a. The equivalence classes defined by an equivalence relation on the set A subdivide A in the following sense:

Proposition 6.5. *Assume we are given an equivalence relation on a set* A*. For all* $a_1, a_2 \in A$, $[a_1] = [a_2]$ *or* $[a_1] \cap [a_2] = \varnothing$.

A **partition** of A is a set Π consisting of subsets of A such that whenever $P_1, P_2 \in \Pi$ with $P_1 \neq P_2$, we have

$$P_1 \cap P_2 = \varnothing,$$

and every $a \in A$ belongs to some $P \in \Pi$.

You may think of Π *as dividing the set A into a collection of disjoint subsets, like fields on a farm. For example, our birthday and first-name equivalence relations give partitions of the club.*

Proposition 6.6. *Given an equivalence relation on* A*, its equivalence classes form a partition of* A*. Conversely, given a partition* Π *of* A*, define* $\sim$ *by* $a \sim b$ *if and only if* a *and* b *lie in the same element of* Π*. Then* $\sim$ *is an equivalence relation.*

The **absolute value** of an integer x is defined as

$$|x| = \begin{cases} x & \text{if } x \geq 0, \\ -x & \text{if } x < 0. \end{cases}$$

We will discuss absolute value in more detail in Section 10.2.

Project 6.7. For each of the following relations defined on $\mathbf{Z}$, determine whether it is an equivalence relation. If it is, determine the equivalence classes.

(i) $x \sim y$ if $x < y$.

(ii) $x \sim y$ if $x \leq y$.

(iii) $x \sim y$ if $|x| = |y|$.

(iv) $x \sim y$ if $x \neq y$.

(v) $x \sim y$ if $xy > 0$.

(vi) $x \sim y$ if $x \mid y$ or $y \mid x$.

Recall that $x \mid y$ means that $\exists k \in \mathbf{Z}$ such that $y = kx$.

Project 6.8. Prove that each of the following relations defined on $\mathbf{Z} \times \mathbf{Z}$ is an equivalence relation. Determine the equivalence classes for each relation.

(i) $(x,y) \sim (v,w)$ if $x^2 + y^2 = v^2 + w^2$.

(ii) $(x,y) \sim (v,w)$ if $y - x^2 = w - v^2$.

(iii) $(x,y) \sim (v,w)$ if $xy = vw$.

(iv) $(x,y) \sim (v,w)$ if $x + 2y = v + 2w$.

This project has a connection to Section 11.1.

Project 6.9. On $\mathbf{Z} \times (\mathbf{Z} - \{0\})$ we define the relation $(m_1,n_1) \sim (m_2,n_2)$ if $m_1 n_2 = n_1 m_2$.

(i) Show that this is an equivalence relation.

(ii) For two equivalence classes $[(m_1,n_1)]$ and $[(m_2,n_2)]$, we define two binary operations $\oplus$ and $\odot$ via

$$[(m_1,n_1)] \oplus [(m_2,n_2)] = [(m_1 n_2 + m_2 n_1, n_1 n_2)]$$

and

$$[(m_1,n_1)] \odot [(m_2,n_2)] = [(m_1 m_2, n_1 n_2)] .$$

What properties do the binary operations $\oplus$ and $\odot$ have?

Project 6.10. High-school geometry is about figures in the plane. One topic you have studied is the idea of two triangles being similar. Prove that the similarity relation is an equivalence relation on the set of all triangles, and describe the equivalence classes.

Here is one more example of an equivalence relation that you might be familiar with from linear algebra.

Example 6.11. Let V be a finite-dimensional vector space and let W be a linear subspace. Define the equivalence relation $\sim$ on V by $v_1 \sim v_2$ if $v_1 - v_2 \in W$. The set of equivalence classes is denoted by V/W and is again a vector space. In the language of linear algebra, the map $v \mapsto [v]$ is a surjective linear map $V \to V/W$ whose kernel is W.

Project 6.12. Construct equivalence relations in other areas of mathematics.

6.2 The Division Algorithm

In Section 6.3, we will discuss an example of an equivalence relation in detail. For this we need a theorem that expresses something you always knew: that when you divide one positive integer into another, there will be a remainder (possibly 0) that is less than the dividing number.

Theorem 6.13 (Division Algorithm). *Let $n \in \mathbf{N}$. For every $m \in \mathbf{Z}$ there exist unique $q, r \in \mathbf{Z}$ such that*

$$m = qn + r \qquad \text{and} \qquad 0 \le r \le n-1 .$$

We call q the **quotient** and r the **remainder** when dividing n into m.

The word algorithm is slightly misleading here: the Division Algorithm is a theorem, not an algorithm. Hint for a proof: Consider first the case $m \ge 0$ by induction on m, and then the case $m < 0$.

Example 6.14. Here are a few instances of the Division Algorithm at work. We think of n and m as given, and of q and r as outputs of the Division Algorithm. For example:

- For $n = 2$ and $m = 9$, we obtain $q = 4$ and $r = 1$.
- For $n = 4$ and $m = 34$, we obtain $q = 8$ and $r = 2$.
- For $n = 5$ and $m = 45$, we obtain $q = 9$ and $r = 0$.
- For $n = 7$ and $m = -16$, we obtain $q = -3$ and $r = 5$.

Proposition 6.15. *The integer m is odd if and only if there exists $q \in \mathbf{Z}$ such that $m = 2q + 1$.*

Proposition 6.16. *For every $n \in \mathbf{Z}$, n is even or $n+1$ is even.*

Proposition 6.17. *Let $m \in \mathbf{Z}$. This number m is even if and only if m^2 is even.*

A **(integer) polynomial** is an expression of the form

$$p(x) = a_d x^d + a_{d-1} x^{d-1} + \cdots + a_1 x + a_0 ,$$

where $a_0, a_1, \ldots, a_d$ are integers (the **coefficients** of the polynomial); we think of x as a variable. Assuming $a_d \neq 0$, we call d the **degree** of $p(x)$. The **zero polynomial** is a polynomial all of whose coefficients are 0.

For real polynomials, we substitute each occurrence of "integer" by "real" (which we may do after Chapter 8). All definitions and propositions here make sense for real polynomials.

Proposition 6.18 (Division Algorithm for Polynomials). *Let $n(x)$ be a polynomial that is not zero. For every polynomial $m(x)$, there exist polynomials $q(x)$ and $r(x)$ such that*

$$m(x) = q(x)\, n(x) + r(x)$$

and either $r(x)$ is zero or the degree of $r(x)$ is smaller than the degree of $n(x)$.

Hint: proof by induction on d, the degree of $m(x) = a_d x^d + \cdots + a_0$. If d is larger than the degree of $n(x) = b_e x^e + \cdots + b_0$, use the induction hypothesis on $m(x) - \frac{a_d}{b_e} x^{d-e} n(x)$.

A **root** of the polynomial $p(x) = a_d x^d + a_{d-1} x^{d-1} + \cdots + a_1 x + a_0$ is a number z such that the polynomial evaluated at z is zero, that is,

$$p(z) = a_d z^d + a_{d-1} z^{d-1} + \cdots + a_1 z + a_0 = 0 .$$

Proposition 6.19. *Let $p(x)$ be a polynomial. The number z is a root of $p(x)$ if and only if there exists a polynomial $q(x)$ such that*

$$p(x) = (x - z)\, q(x) .$$

Proposition 6.20. *A polynomial of degree d has at most d roots.*

6.3 The Integers Modulo n

In this section, we discuss in detail an example of an equivalence relation, namely, "clock arithmetic" (except that we do not require our clocks to have 12 numbers). Given a fixed $n \in \mathbf{N}$, we define the relation $\equiv$ on $\mathbf{Z}$ by

This definition makes sense but is not useful when $n = 1$, so for practical purposes we may assume that $n \geq 2$.

$$x \equiv y \qquad \text{if} \qquad x - y \text{ is divisible by } n.$$

When there is any possibility of ambiguity about n (the **modulus**), we write this as

$$x \equiv y \pmod{n} .$$

Example 6.21. We discuss the case $n = 2$. Here $x \equiv y \pmod 2$ means that $x - y$ is even, i.e., x and y are either both even or both odd. This is what is meant by saying that x and y have the **same parity**.

Project 6.22. Discuss in what ways the relation $\equiv$ (for different n) generalizes the notion of parity.

Example 6.23. Given $n \in \mathbf{N}$, every integer m has a quotient q and a remainder r when divided by n, by the Division Algorithm (Theorem 6.13). Then $m \equiv r \pmod{n}$.

Proposition 6.24. *Fix a modulus $n \in \mathbf{N}$.*

(i) *$\equiv$ is an equivalence relation on $\mathbf{Z}$.*

(ii) *The equivalence relation $\equiv$ has exactly n distinct equivalence classes, namely $[0], [1], \ldots, [n-1]$.*

The set of equivalence classes is called the **set of integers modulo** n, written $\mathbf{Z}_n$ or $\mathbf{Z}/n\mathbf{Z}$.

Proof. (i) We need to check that $\equiv$ satisfies reflexivity, symmetry, and transitivity.

$a \equiv a$ means that n divides $a - a = 0$, which is certainly true.

Assume $a \equiv b$, i.e., n divides $a - b$. Then there exists $j \in \mathbf{Z}$ such that

$$a - b = jn .$$

But then

$$b - a = -jn$$

is also divisible by n, i.e., $b \equiv a$.

Assume $a \equiv b$ and $b \equiv c$, i.e., n divides both $a - b$ and $b - c$. This means that there exist integers j and k such that

$$a - b = jn \qquad \text{and} \qquad b - c = kn .$$

But then

$$a - c = (a - b) + (b - c) = jn + kn = (j + k)n$$

is divisible by n, i.e., $a \equiv c$.

(ii) We need to prove that every integer falls into one of the equivalence classes $[0], [1], \ldots, [n-1]$, and that they are all distinct.

For each $m \in \mathbf{Z}$, we can, by the Division Algorithm (Theorem 6.13), find integers q, r with $0 \le r \le n - 1$ such that $m = qn + r$. In particular, $m - r$ is divisible by n. But then $m \equiv r$, and by Proposition 6.4, $[m] = [r]$, which is one of the equivalence classes $[0], [1], \ldots, [n-1]$.

Now assume $0 \le m, k \le n - 1$ and $[m] = [k]$. Our goal is to show that $m = k$. Again by Proposition 6.4, $[m] = [k]$ is equivalent to $m \equiv k$, i.e., n divides $m - k$. But by construction,

$$-n + 1 \le m - k \le n - 1 ,$$

and the only number divisible by n in this range is 0, that is, $m = k$. □

Proposition 6.25. *If $a \equiv a' \pmod{n}$ and $b \equiv b' \pmod{n}$ then*

$$a + b \equiv a' + b' \pmod{n} \qquad \textit{and} \qquad ab \equiv a'b' \pmod{n} .$$

This proposition allows us to make the following definition: For elements $[a]$ and $[b]$ of $\mathbf{Z}_n$, we define **addition** $\oplus$ and **multiplication** $\odot$ on $\mathbf{Z}_n$ via

Here $\oplus$ and $\odot$ happen in $\mathbf{Z}_n$, whereas $+$ and $\cdot$ happen in $\mathbf{Z}$.

$$[a] \oplus [b] = [a + b] \qquad \text{and} \qquad [a] \odot [b] = [ab] .$$

Proposition 6.26. *Fix an integer $n \ge 2$. Addition $\oplus$ and multiplication $\odot$ on $\mathbf{Z}_n$ are commutative, associative, and distributive. The set $\mathbf{Z}_n$ has an additive identity, a multiplicative identity, and additive inverses.*

Proposition 6.26 says that Axioms 1.1–1.4 hold in $\mathbf{Z}_n$.

Project 6.27. Study for which n the set $\mathbf{Z}_n$ satisfies the cancellation property (Axiom 1.5). Prove your assertions.

The set $\mathbf{Z}_n$ is of fundamental importance in mathematics. For example, many computer encryption schemes are based on arithmetic in $\mathbf{Z}_n$; we will give an example in Chapter B. Among the different $\mathbf{Z}_n$'s those for which n is prime are particularly useful, as we will see in the next section.

6.4 Prime Numbers

A prime integer is also called a prime number *or simply a* prime.

An integer $n \geq 2$ is **prime** if it is divisible by only ± 1 and $\pm n$. The first sixteen primes are 2, 3, 5, 7, 11, 13, 17, 19, 23, 29, 31, 37, 41, 43, 47, 53. An integer ≥ 2 that is not prime is called **composite**. If we can write $n = q_1 q_2 \cdots q_k$ then the numbers $q_1, q_2, \ldots, q_k$ are **factors** of n, and this product is a **factorization** of n.

Proposition 6.28 will be substantially strengthened in Theorem 6.32.

Proposition 6.28. *Every integer* ≥ 2 *can be factored into primes.*

Proposition 6.28 is easy to prove, for example by induction. Unfortunately, it does not suffice for most purposes: we often need the fact that such a **prime factorization** is *unique* (except for reordering of the primes in the factorization). Our next goal is to prove just that.

We need the Well-Ordering Principle (Theorem 2.32) here to ensure that this definition makes sense.

Along the way, recall from Section 2.4 that $\gcd(m,n)$ is the smallest element of the set

$$S = \{k \in \mathbf{N} : k = mx + ny \text{ for some } x, y \in \mathbf{Z}\} .$$

This set is empty when $m = n = 0$, in which case we define $\gcd(0,0) = 0$.

Proposition 6.29. *Let* $m, n \in \mathbf{Z}$.

(i) $\gcd(m,n)$ *divides both* m *and* n.

(ii) *Unless* m *and* n *are both* 0, $\gcd(m,n) > 0$.

(iii) *Every integer that divides both* m *and* n *also divides* $\gcd(m,n)$.

Together with Proposition 2.23, Proposition 6.29 implies that $\gcd(m,n)$ is the largest integer that divides both m and n. This finally explains our notation: $\gcd(m,n)$ is called the **greatest common divisor** of m and n.

Proof of Proposition 6.29. Let $g = \gcd(m,n)$, i.e., g is the smallest element of

$$S = \{k \in \mathbf{N} : k = mx + ny \text{ for some } x, y \in \mathbf{Z}\} .$$

If $m = n = 0$ then $g = 0$ and the statement of Proposition 6.29 holds.

If $m = 0$ and $n \neq 0$ then

$$S = \{|n|y : y \in \mathbf{N}\}$$

and $g = |n|$, which satisfies the three properties in the proposition. The case $m \neq 0$, $n = 0$ is analogous.

Now assume that neither m nor n is zero. Then S will remain unchanged if we switch m with $-m$ (or n with $-n$); so we may assume that both m and n are positive.

(i) Suppose (by way of contradiction) that g does not divide m. By the Division Algorithm (Theorem 6.13), there exist $q, r \in \mathbf{Z}$ such that

$$m = qg + r \qquad \text{and} \qquad 0 < r < g$$

(if $r = 0$ then g would divide m). By definition, $g = mx + ny$ for some $x, y \in \mathbf{Z}$, and so

$$r = m - qg = m - q(mx + ny) = m(1 - qx) + n(-qy),$$

which implies that $r \in S$. But $0 < r < g$, which contradicts the fact that g is the smallest element of S.

(ii) This follows from the definition of the greatest common divisor.

(iii) Assume a divides both m and n. Then $m = j_1 a$ and $n = j_2 a$ for some $j_1, j_2 \in \mathbf{Z}$. But then we have for each $x, y \in \mathbf{Z}$,

$$mx + ny = (j_1 x + j_2 y)\, a,$$

that is, a divides every element of S; in particular, a divides g. □

Proposition 6.30. *For all $k, m, n \in \mathbf{Z}$*

$$\gcd(km, kn) = |k| \gcd(m, n).$$

Proposition 6.31 (Euclid's lemma). *Let p be prime and $m, n \in \mathbf{N}$. If $p \mid mn$ then $p \mid m$ or $p \mid n$.*

Hint: If p does not divide m, then $\gcd(p, m) = 1$. Warning: it is tempting to use Theorem 6.32 to prove Euclid's lemma, but we need Euclid's lemma to prove Theorem 6.32.

Theorem 6.32. *Every integer ≥ 2 can be factored uniquely into primes.*

Here "unique" means "unique up to ordering": for example, $12 = 2^2 \cdot 3 = 2 \cdot 3 \cdot 2 = 3 \cdot 2^2$.

Proposition 6.33. *Let $m, n \in \mathbf{N}$. If m divides n and p is a prime factor of n that is not a prime factor of m, then m divides $\frac{n}{p}$.*

Proposition 6.34. *If p is prime and $0 < r < p$ then $\binom{p}{r}$ is divisible by p.*

Proof. Assume p is prime and $0 < r < p$. Then none of the numbers $2, 3, \ldots, r$ and $2, 3, \ldots, p - r$ divides p. On the other hand, we know that

$$\binom{p}{r} = \frac{p!}{r!(p-r)!} = \frac{p \cdot (p-1)!}{r!(p-r)!}$$

is an integer. By Proposition 6.33, since $2,3,\ldots,r$ and $2,3,\ldots,p-r$ do not divide p, they have to divide $(p-1)!$, i.e., $\frac{(p-1)!}{r!(p-r)!}$ is an integer. But then $\binom{p}{r} = p\,\frac{(p-1)!}{r!(p-r)!}$ implies that p divides $\binom{p}{r}$. □

Hint: this is a lovely application of the Binomial Theorem 4.21.

Theorem 6.35 (Fermat's little theorem). *If $m \in \mathbf{Z}$ and p is prime, then*

$$m^p \equiv m \pmod{p}.$$

Corollary 6.36. *Let $m \in \mathbf{Z}$ and let p be a prime that does not divide m. Then*

$$m^{p-1} \equiv 1 \pmod{p}.$$

This open problem raises the much simpler question whether there are infinitely many primes. The answer is not entirely obvious, is yes, and has been known for at least 2300 years. Prove it!

Over the centuries, people have found that while the definition of a prime is easy, it is difficult to understand how the primes are distributed among the natural numbers. For example, if two primes differ by 2 (i.e., p and $p+2$ are both prime) then, as a pair, they are called **twin primes**. Examples are $(3,5)$, $(17,19)$, and $(41,43)$. It is unknown whether there are infinitely many pairs of twin primes or whether there are only finitely many. If it turns out that there are only finitely many, the exact number of pairs of twin primes would be an intriguing number, since it would measure something fundamental.

Project 6.37. Many books on number theory use the statement of Proposition 6.29 as the definition of the greatest common divisor. But then the authors have to prove that the greatest common divisor of two numbers always exists. Think about how this could be done.

Review Questions. Do you understand the set $\mathbf{Z}_n$ of integers modulo n? Do you see that this is an example of a set of equivalence classes for which the corresponding equivalence relation is on the set $\mathbf{Z}$ of integers?

Weekly reminder: Reading mathematics is not like reading novels or history. You need to think slowly about every sentence. Usually, you will need to reread the same material later, often more than one rereading.

This is a short book. Its core material occupies about 140 pages. Yet it takes a semester for most students to master this material. In summary: read line by line, not page by page.

Chapter 7
Arithmetic in Base Ten

If people do not believe that mathematics is simple, it is only because they do not realize how complicated life is.
John von Neumann (1903–1957)

Before You Get Started. The whole literate world has been taught that every nonnegative integer can be represented by a finite string of digits, and that different strings of digits correspond to different integers. None of this is in our axioms, so it must be established. You know that the string 365 means $5+6\cdot 10+3\cdot 100$ and you know that the string 371 means $1+7\cdot 10+3\cdot 100$. These sums add up to different integers. Are you sure? How do you know? Are you equally sure when you have two strings of 400 digits that are not exactly the same? And while we are questioning basic things, here is another problem: In elementary school you learned how to add strings of integers like 365 and 371. How did you do it? And why does it work? Can you write down the instructions so that someone could add other numbers? How did your elementary-school teacher explain addition to you?

M. Beck and R. Geoghegan, *The Art of Proof: Basic Training for Deeper Mathematics*, Undergraduate Texts in Mathematics, DOI 10.1007/978-1-4419-7023-7_7,

7.1 Base-Ten Representation of Integers

In our axioms two (distinct!) elements of $\mathbf{Z}$ were given names: 0 and 1. Later some more integers were given names: 2, 3, 4, 5, 6, 7, 8, 9. Now we give the name 10 to the integer $9+1$.

Proposition 7.1. *If $n \in \mathbf{N}$ then $n < 10^n$.*

In the language of Section 5.4, $\nu : \mathbf{Z}_{\geq 0} \to \mathbf{N}$ is a function.

Define $\nu(0) = 1$, and for all $n \in \mathbf{N}$, let $\nu(n)$ be the smallest element of

$$\left\{t \in \mathbf{N} : n < 10^t\right\}.$$

The number $\nu(n)$ is called **the number of digits of n with respect to base 10**. Our definition of ν makes sense because, by Proposition 7.1, $\{t \in \mathbf{N} : n < 10^t\}$ contains n and is therefore nonempty, so the Well-Ordering Principle (Theorem 2.32) guarantees that this set contains a unique smallest element.

Example 7.2. $\nu(d) = 1$ for all digits d, and $\nu(10) = 2$.

Proposition 7.3. *For all $n \in \mathbf{N}$, $\nu(n) = k$ if and only if $10^{k-1} \leq n < 10^k$.*

Corollary 7.4. *If $\nu(n) > \nu(n-1)$ then n is a power of* 10.

The Division Algorithm (Theorem 6.13) is being used here.

Proposition 7.5. *Given $n \in \mathbf{N}$, write $n = 10q + r$, where $q, r \in \mathbf{Z}$ and $0 \leq r \leq 9$. Then $\nu(n) = \nu(q) + 1$.*

Corollary 7.6. *Given $n \in \mathbf{N}$, write $n = 10q + r$, where $q, r \in \mathbf{Z}$ and $0 \leq r \leq 9$. Then $q < n$.*

Theorem 7.7 (Existence of base-ten representation for positive integers). *Let $n \in \mathbf{N}$. Then there exist digits $x_0, x_1, \ldots, x_{\nu(n)-1}$ such that*

$$n = \sum_{i=0}^{\nu(n)-1} x_i \, 10^i.$$

and $x_{\nu(n)-1} > 0$.

Proof. We prove this theorem by induction on n. For the base case $n = 1$, we have $\nu(1) = 1$, and thus $n = 1 = \sum_{i=0}^{0} x_i \, 10^i$ with $x_0 = 1$.

For the induction step, assume that the statement of Theorem 7.7 is true whenever n is replaced by an integer smaller than n. The Division Algorithm (Theorem 6.13) says that $n = 10q + r$, where $q, r \in \mathbf{Z}$ and $0 \leq r \leq 9$. By the induction hypothesis and Proposition 7.5, we can write

$$q = \sum_{i=0}^{\nu(n)-2} y_i \, 10^i ,$$

Corollary 7.6 allows us to use the induction hypothesis for q.

where the y_i's are digits. But then

$$n = 10q + r = 10 \sum_{i=0}^{\nu(n)-2} y_i \, 10^i + r = \sum_{i=0}^{\nu(n)-2} y_i \, 10^{i+1} + r = \sum_{i=0}^{\nu(n)-1} x_i \, 10^i ,$$

where $x_0 = r$ and $x_i = y_{i-1}$ for $1 \le i \le \nu(n) - 1$. □

Proposition 7.8. *For all* $r \in \mathbf{N}$, $\left(\sum_{i=0}^{r-1} 9 \cdot 10^i \right) + 1 = 10^r$.

Proposition 7.9 (Uniqueness of base-ten representation for positive integers). *Let* $n \in \mathbf{N}$. *If*

$$n = \sum_{i=0}^{p} x_i 10^i = \sum_{i=0}^{q} y_i 10^i ,$$

where $p, q \in \mathbf{Z}_{\ge 0}$, *each* x_i *and each* y_i *is a digit,* $x_p \ne 0$, *and* $y_q \ne 0$, *then* $p = q$, *and* $x_i = y_i$ *for all* i.

Proposition 7.10. *Let* $n \in \mathbf{N}$. *Then* n *is divisible by* 3 *if and only if the sum of its digits is divisible by* 3.

Hint: Write $n = \sum_{i=0}^{\nu(n)-1} x_i \, 10^i$ as in Theorem 7.7. Define $\sigma(n) = \sum_{i=0}^{\nu(n)-1} x_i$. *Start by proving that* $n - \sigma(n)$ *is divisible by* 3.

One approach to the last proposition is through modular arithmetic. In fact, the following statement is a generalization of Proposition 7.10 (and you should prove that Proposition 7.10 follows from Proposition 7.11):

Proposition 7.11. *Let* $n = \sum_{i=0}^{\nu(n)-1} x_i \, 10^i$, *where each* x_i *is a digit; then*

$$n \equiv x_0 + x_1 + x_2 + \cdots + x_{\nu(n)-1} \pmod{3} .$$

As you might imagine, the "divisibility test by 3" given in Proposition 7.10 and the related modular identity in Proposition 7.11 are not the end of the story. Here is a small sample of similar modular identities.

Proposition 7.12. *Let* $n = \sum_{i=0}^{\nu(n)-1} x_i \, 10^i$, *where each* x_i *is a digit.*

(i) $n \equiv x_0 \pmod{2}$.

(ii) $n \equiv x_0 + 10x_1 \pmod{4}$.

(iii) $n \equiv x_0 + 10x_1 + 100x_2 \pmod{8}$.

(iv) $n \equiv x_0 \pmod{5}$.

(v) $n \equiv x_0 \pmod{10}$.

(vi) $n \equiv x_0 + x_1 + x_2 + \cdots \pmod{9}$.

(vii) $n \equiv x_0 - x_1 + x_2 - \cdots \pmod{11}$.

(viii) $n \equiv (x_0 + 10x_1 + 100x_2) - (x_3 + 10x_4 + 100x_5) + (x_6 + 10x_7 + 100x_8) - \cdots \pmod{7}$.

(ix) $n \equiv (x_0 + 10x_1 + 100x_2) - (x_3 + 10x_4 + 100x_5) + (x_6 + 10x_7 + 100x_8) - \cdots \pmod{11}$.

(x) $n \equiv (x_0 + 10x_1 + 100x_2) - (x_3 + 10x_4 + 100x_5) + (x_6 + 10x_7 + 100x_8) - \cdots \pmod{13}$.

(xi) $n \equiv (x_0 + 3x_1 + 2x_2) - (x_3 + 3x_4 + 2x_5) + (x_6 + 3x_7 + 2x_8) - \cdots \pmod{7}$.

Project 7.13. Each part in Proposition 7.12 gives rise to a divisibility test as in Proposition 7.10. State and prove these divisibility tests.

This test was discovered by Apoorva Khare when she was a high-school senior in Orissa, India; see Electronic Journal of Undergraduate Mathematics **3** *(1997), 1–5.*

Project 7.14. Prove the following divisibility test: $n = \sum_{i=0}^{\nu(n)-1} x_i \, 10^i$ is divisible by 7 if and only if

$$(-2)^{\nu(n)-1} x_0 + (-2)^{\nu(n)-2} x_1 + (-2)^{\nu(n)-3} x_2 + \cdots + (-2) x_{\nu(n)-2} + x_{\nu(n)-1}$$

is divisible by 7. Generalize.

Let $m \in \mathbf{Z}_{\geq 0}$. By Theorem 7.7 and Proposition 7.9, for each $i \in \mathbf{Z}_{\geq 0}$ such that $0 \leq i \leq \nu(m) - 1$ there is a unique digit x_i such that $m = \sum_{i=0}^{\nu(m)-1} x_i \, 10^i$ and $x_{\nu(m)-1} > 0$. It is convenient (as we have all been taught since childhood) to represent m by the string of digits $x_{\nu(m)-1} x_{\nu(m)-2} \cdots x_2 x_1 x_0$. For example,

$$m = 3 \cdot 10^0 + 6 \cdot 10^1 + 8 \cdot 10^2 + 0 \cdot 10^3 + 7 \cdot 10^4$$

is represented by 70863. This string is called the **base-ten representation** of m.

The following is a criterion for deciding when $m < n$:

To prove Theorem 7.15, one way to proceed is:

(a) $9 \sum_{i=0}^{k} 10^i < 10^{k+1}$.
(b) *If* $m < n$ *then* $\nu(m) \leq \nu(n)$.
(c) *If* $m < n$ *either* (i) *or* (ii) *holds.*
(d) *If* (i) *or* (ii) *holds, then* $m < n$.

Theorem 7.15. *Let* $m, n \in \mathbf{Z}_{\geq 0}$. *Assume m and n have the base-ten representations* $x_{\nu(m)-1} x_{\nu(m)-2} \cdots x_2 x_1 x_0$ *and* $y_{\nu(n)-1} y_{\nu(n)-2} \cdots y_2 y_1 y_0$, *respectively. Then* $m < n$ *if and only if either*

(i) $\nu(m) < \nu(n)$

or

(ii) $\nu(m) = \nu(n)$ *and* $x_j < y_j$, *where j is the smallest element of*

$$\{ i \in \mathbf{Z}_{\geq 0} : x_k = y_k \text{ for all } k > i \} .$$

Other Bases. Every integer $k \geq 2$ can be used as a base for representing the integers. Base 2 (binary), base 8 (octal), and base 16 (hexadecimal) are used in computer code. This section has been written so that the proof of existence and uniqueness can easily be adapted from the case $k = 10$ by making simple changes as follows:

Why should we not use base 1?

- The definition of *digit* must be changed. For base k, the digits are $0, 1, \ldots, k-1$, but we might use different symbols. For example, for base 2 use 0 and 1. For base 12 we use 0, 1, 2, 3, 4, 5, 6, 7, 8, 9, a, b.
- In the definition of $\nu(n)$ replace 10 by k.
- In the proofs of the existence and uniqueness theorems replace 10 by k everywhere, and use the new ν.

Project 7.16. Describe a mathematical procedure (an algorithm) that converts the decimal (base-10) description of an integer to its octal (base-8) description.

You might have noticed that in this chapter we deal only with base-10 representations of *positive* integers. However, as you well know, the base-10 representation of a negative integer n is simply that of the positive integer $-n$ preceded by a minus sign. (The base-10 representation of 0 is 0.)

7.2 The Addition Algorithm for Two Nonnegative Numbers (Base 10)

An **algorithm** is a procedure for doing something mathematical, step by step.

That is not a formal definition: it is not an easy matter to write down the formal meaning of the word algorithm.

We saw that each $m \in \mathbf{Z}_{\geq 0}$ has a unique representation

$$m = \sum_{i=0}^{\nu(m)-1} x_i \, 10^i.$$

If $n \in \mathbf{Z}_{\geq 0}$ and

$$n = \sum_{i=0}^{\nu(n)-1} y_i \, 10^i,$$

we want an algorithm for the digits $z_0, z_1, \ldots$ when $m+n$ is written as

$$m+n = \sum_{i=0}^{\nu(m+n)-1} z_i \, 10^i.$$

We are all familiar with this: for example, if $m = 332$ and $n = 841$ your previous knowledge of mathematics leads you to believe the statement $m+n = 1173$. What did you do to get 1173? Our goal is to describe the process rigorously.

Note that we allow $x_q = 0$ or $y_q = 0$, so m and n might not have standard base-ten representations. The reason for doing it this way is easily seen if you add 27 to 4641. We are in effect adding 0027 to 4641.

The Algorithm: Given as input digits $x_0, \dots, x_q$ and $y_0, \dots, y_q$, the output of the algorithm consists of ordered pairs $(z_0, i_0), \dots, (z_{q+1}, i_{q+1})$, where each z_k is a digit and each i_k is 0 or 1. The z_k's and i_k's are defined in stages recursively:

Stage 0: The input consists of two digits x_0 and y_0. The output is (z_0, i_0) as follows: if $x_0 + y_0 < 10$ then $z_0 = x_0 + y_0$ and $i_0 = 0$; if $10 \le x_0 + y_0$ then $x_0 + y_0 = d_0 + 10$, where d_0 is a digit (why?) and in that case $z_0 = d_0$ and $i_0 = 1$.

Stage k: Here, $1 \le k \le q$. The input consists of two digits x_k and y_k as well as i_{k-1}, which is either 0 or 1 and which has been found in Stage $k-1$. The output is (z_k, i_k) as follows:

Case 1: If $x_k + y_k < 10$ and $i_{k-1} = 0$, define $z_k = x_k + y_k$ and $i_k = 0$.

Case 2: If $x_k + y_k < 9$ and $i_{k-1} = 1$, define $z_k = x_k + y_k + 1$ and $i_k = 0$.

Case 3: If $10 \le x_k + y_k$ and $i_{k-1} = 0$ then there is a unique digit d_k such that $x_k + y_k = d_k + 10$; define $z_k = d_k$ and $i_k = 1$.

Case 4: If $9 \le x_k + y_k$ and $i_{k-1} = 1$ then there is a unique digit d_k such that $x_k + y_k + 1 = d_k + 10$; define $z_k = d_k$ and $i_k = 1$.

Stage $q+1$: Define $z_{q+1} = i_q$ and $i_{q+1} = 0$.

We remark that the output (z_0, i_0) depends only on x_0 and y_0. If $k \ge 1$, the output (z_k, i_k) depends only on x_k, y_k and i_{k-1}.

You probably believe that this is the algorithm you were taught in elementary school. But does it give the correct answer? That is the subject of the next theorem.

Theorem 7.17. *If $m = \sum_{i=0}^{q} x_i\, 10^i$ and $n = \sum_{i=0}^{q} y_i\, 10^i$, where each x_i and each y_i is a digit, then $m + n = \sum_{i=0}^{q+1} z_i\, 10^i$, where the digits $z_0, \dots, z_{q+1}$ are obtained from the algorithm.*

Proof. We proceed by induction on q. In the base case $q = 0$, the theorem says that $x_0 + y_0 = z_0 + i_0 \cdot 10$, which is true—just look at Stage 0.

For the induction step, assume the theorem is true whenever q is replaced by $q-1$. When $x_0, \dots, x_{q-1}$ and $y_0, \dots, y_{q-1}$ are the input, let $(z'_0, i'_0), \dots, (z'_q, i'_q)$ form the output. By the induction hypothesis,

$$\sum_{i=0}^{q-1} x_i\, 10^i + \sum_{i=0}^{q-1} y_i\, 10^i = \sum_{i=0}^{q} z_i\, 10^i.$$

We already remarked that $z'_k = z_k$ when $k \leq q-1$. The last stage of the algorithm gives $z'_q = i_{q-1}$. So

$$\begin{aligned} m+n &= \sum_{i=0}^{q-1} x_i\, 10^i + x_q\, 10^q + \sum_{i=0}^{q-1} y_i\, 10^i + y_q\, 10^q \\ &= \sum_{i=0}^{q-1} z_i\, 10^i + i_{q-1}\, 10^q + x_q\, 10^q + y_q\, 10^q \\ &= \sum_{i=0}^{q+1} z_i\, 10^i, \end{aligned}$$

since the algorithm gives $z_q\, 10^q + z_{q+1}\, 10^{q+1} = (x_q + y_q + i_{q-1})10^q$. □

Proposition 7.18. *Let p be the maximum of the numbers $\nu(m)$ and $\nu(n)$. Then $\nu(m+n) = p$ or $p+1$.*

Hint: Use Exercise 7.3.

Example 7.19. (i) If $m = 332$, $n = 841$, then $p = 3$ and $\nu(m+n) = 4$.

(ii) If $m = 32$, $n = 641$, then $p = 3$ and $\nu(m+n) = 3$.

Proposition 7.20. *Using the notation of the algorithm on the previous page, if x_q and y_q are not both zero, then $z_q = 0$ or 1. If $z_q = 0$ then $\nu(m+n) = q$. If $z_q = 1$ then $\nu(m+n) = q+1$.*

Proof. By the last stage of the algorithm, $z_q = i_{q-1}$, which is 0 or 1. If $z_q = 1$ then $\nu(m+n)$ is $q+1$. If $z_q = 0$ then $\nu(m+n) \leq q$. By Proposition 7.18, $\nu(m+n) = q$. □

The same approach can be used to prove the correctness of the other elementary-school algorithms:

(i) Subtraction

$$\begin{array}{r} 461 \\ 29 \\ \hline 432 \end{array}$$

(ii) Long addition

$$\begin{array}{r} 461 \\ 29 \\ 391 \\ \hline 881 \end{array}$$

(iii) Long multiplication

$$\begin{array}{r} 461 \\ 29 \\ \hline 4149 \\ 9220 \\ \hline 13369 \end{array}$$

More about this problem can be found in John Holte's article Carries, combinatorics, and an amazing matrix, *American Mathematical Monthly* **104** *(1997), 138–149.*

Project 7.21. In our algorithm for adding base-ten representations of integers, we implicitly introduced the action of "carrying" when adding digits whose sum is larger than 9 (the i_k's are the "carries"). Randomly choose 200 digits and use them to make up two 100-digit numbers. If you add these two numbers, how often do you expect to "carry"? How about if you add three "random" 100-digit numbers? Or four? Experiment.

Review Questions. Do you understand what an algorithm is? And that the procedure for addition of two numbers that you learned in elementary school is a nice example of an algorithm? Do you see why it is necessary to write down an algorithm in a careful and formal way, in order (for example) to write a computer program?

Weekly reminder: Reading mathematics is not like reading novels or history. You need to think slowly about every sentence. Usually, you will need to reread the same material later, often more than one rereading.

This is a short book. Its core material occupies about 140 pages. Yet it takes a semester for most students to master this material. In summary: read line by line, not page by page.

Part II: The Continuous

Chapter 8
Real Numbers

Mathematical study and research are very suggestive of mountaineering. Whymper made several efforts before he climbed the Matterhorn in the 1860's and even then it cost the life of four of his party. Now, however, any tourist can be hauled up for a small cost, and perhaps does not appreciate the difficulty of the original ascent. So in mathematics, it may be found hard to realise the great initial difficulty of making a little step which now seems so natural and obvious, and it may not be surprising if such a step has been found and lost again.
Louis Joel Mordell (1888–1972)

Before You Get Started. Just like the integers, the real numbers, which ought to include the integers but also numbers like $\frac{1}{3}$, $-\sqrt{2}$, and π, will be defined by a set of axioms. From what you know about real numbers, what should this set of axioms include? How should the axioms differ from those of Chapters 1 and 2?

M. Beck and R. Geoghegan, *The Art of Proof: Basic Training for Deeper Mathematics*, Undergraduate Texts in Mathematics, DOI 10.1007/978-1-4419-7023-7_8,

We start all over again. You have used the real numbers in calculus. You have pictured them as points on an x-axis or a y-axis. You have probably been told that there is a bijection between the set of points on the x-axis and the set of all real numbers. Even if this was not made explicit in your calculus course, it was implied when you gave a real-number label to an arbitrary point on the x-axis, or when you assumed that there is a point on the x-axis for every real number.

Intuitively, you are familiar with many real numbers: examples are $-\sqrt{2}$, π, and $6e$. You probably thought of the integers as examples of real numbers: you calibrated the x-axis by marking two points as "0" and "1", thus defining one unit of length; and, with that calibration, you knew which point on the x-axis should get the label "7" and which should get the label "−4".

The exact relationship between the integers and the real numbers will require careful discussion in Chapter 9.

We are now going to rebuild your knowledge of the real numbers. In the first stage, which is this chapter, we will define the real numbers by means of axioms, just as we did with the integers in Part I. And as we did with the set of integers **Z**, we will assume without proof that a set **R** satisfying our axioms exists.

8.1 Axioms

We assume that there exists a set, denoted by **R**, whose members are called **real numbers**. This set **R** is equipped with binary operations $+$ and $\cdot$ satisfying Axioms 8.1–8.5, 8.26, and 8.52 below.

Axiom 8.1. *For all* $x, y, z \in \mathbf{R}$*:*

(i) $x+y=y+x$.

(ii) $(x+y)+z=x+(y+z)$.

(iii) $x\cdot(y+z)=x\cdot y+x\cdot z$.

(iv) $x\cdot y=y\cdot x$.

(v) $(x\cdot y)\cdot z=x\cdot(y\cdot z)$.

The product $x\cdot y$ is often written xy.

Axiom 8.2. *There exists a real number* 0 *such that for all* $x\in\mathbf{R}$, $x+0=x$.

Axiom 8.3. *There exists a real number* 1 *such that* $1\neq 0$ *and whenever* $x\in\mathbf{R}$, $x\cdot 1=x$.

Axiom 8.4. *For each* $x\in\mathbf{R}$*, there exists a real number, denoted by* $-x$*, such that* $x+(-x)=0$.

Axiom 8.5. *For each $x \in \mathbf{R} - \{0\}$, there exists a real number, denoted by x^{-1}, such that $x \cdot x^{-1} = 1$.*

Proposition 8.6. *For all $x, y \in \mathbf{R} - \{0\}$, $(xy)^{-1} = x^{-1}y^{-1}$.*

Proposition 8.7. *Let $x, y, z \in \mathbf{R}$ and $x \neq 0$. If $xy = xz$ then $y = z$.*

Proof. Assume $x, y, z \in \mathbf{R}$, $x \neq 0$, and $xy = xz$. By Axiom 8.5, there exists x^{-1}, and thus

$$\begin{aligned} x^{-1}(xy) &= x^{-1}(xz) \\ \left(x^{-1}x\right)y &= \left(x^{-1}x\right)z \\ \left(xx^{-1}\right)y &= \left(xx^{-1}\right)z \\ 1 \cdot y &= 1 \cdot z \\ y &= z. \end{aligned}$$

Here we have used Axioms 8.1(v), 8.1(iv), 8.5, and 8.3. □

Proposition 8.7 is the **R**-analogue of Axiom 1.5 for **Z**: the proposition asserts that the cancellation property described in Axiom 1.5 also holds in **R**. And since Axioms 8.1–8.4 are the same as Axioms 1.1–1.4, any proposition we proved about **Z** using only Axioms 1.1–1.5 is also true for **R**, with an identical proof. We will need to refer to some of the real versions of the propositions proved for **Z**; so we state the corresponding propositions for **R** (which again will have the same proof as those for **Z**) in small font.

Proposition 8.8. *If $m, n, p \in \mathbf{R}$ then $(m+n)p = mp + np$.*

Proposition 8.9. *If $m \in \mathbf{R}$, then $0 + m = m$ and $1 \cdot m = m$.*

Proposition 8.10. *Let $m, n, p \in \mathbf{R}$. If $m + n = m + p$, then $n = p$.*

Proposition 8.11. *Let $m, x_1, x_2 \in \mathbf{R}$. If m, x_1, x_2 satisfy the equations $m + x_1 = 0$ and $m + x_2 = 0$, then $x_1 = x_2$.*

Proposition 8.12. *If $m, n, p, q \in \mathbf{R}$ then*

(i) $(m+n)(p+q) = (mp+np) + (mq+nq)$.

(ii) $m + (n + (p+q)) = (m+n) + (p+q) = ((m+n)+p)+q$.

(iii) $m + (n+p) = (p+m) + n$.

(iv) $m(np) = p(mn)$.

(v) $m(n+(p+q)) = (mn+mp) + mq$.

(vi) $(m(n+p))q = (mn)q + m(pq)$.

Proposition 8.13. *Let $x \in \mathbf{R}$. If x has the property that for each $m \in \mathbf{R}$, $m + x = m$, then $x = 0$.*

Proposition 8.14. *Let $x \in \mathbf{R}$. If x has the property that there exists $m \in \mathbf{R}$ such that $m+x=m$, then $x=0$.*

Proposition 8.15. *For all $m \in \mathbf{R}$, $m \cdot 0 = 0 = 0 \cdot m$.*

Proposition 8.16. *Let $x \in \mathbf{R}$. If x has the property that for all $m \in \mathbf{R}$, $mx = m$, then $x = 1$.*

Proposition 8.17. *Let $x \in \mathbf{R}$. If x has the property that for some nonzero $m \in \mathbf{R}$, $mx = m$, then $x = 1$.*

Proposition 8.18. *For all $m,n \in \mathbf{R}$, $(-m)(-n) = mn$.*

Proposition 8.19.

(i) *For all $m \in \mathbf{R}$, $-(-m) = m$.*

(ii) $-0 = 0$.

Proposition 8.20. *Given $m,n \in \mathbf{R}$ there exists one and only one $x \in \mathbf{R}$ such that $m+x=n$.*

Proposition 8.21. *Let $x \in \mathbf{R}$. If $x \cdot x = x$ then $x = 0$ or 1.*

Proposition 8.22. *For all $m,n \in \mathbf{R}$:*

(i) $-(m+n) = (-m)+(-n)$.

(ii) $-m = (-1)m$.

(iii) $(-m)n = m(-n) = -(mn)$.

Proposition 8.23. *Let $m,n \in \mathbf{R}$. If $mn = 0$, then $m = 0$ or $n = 0$.*

As with $\mathbf{Z}$, we define **subtraction** in $\mathbf{R}$ by

$$x - y := x + (-y).$$

Proposition 8.24. *For all $m,n,p,q \in \mathbf{R}$:*

(i) $(m-n)+(p-q) = (m+p)-(n+q)$.

(ii) $(m-n)-(p-q) = (m+q)-(n+p)$.

(iii) $(m-n)(p-q) = (mp+nq)-(mq+np)$.

(iv) $m-n = p-q$ *if and only if* $m+q = n+p$.

(v) $(m-n)p = mp-np$.

Alternative notations for $\frac{y}{x}$ are y/x and $y \div x$. Do not confuse the division symbol / with the symbol | which describes the divisibility property of integers introduced in Section 1.2.

Here is a definition that we could not make in $\mathbf{Z}$: We define a new operation on $\mathbf{R}$ called **division** by

$$\frac{y}{x} := y \cdot x^{-1}.$$

Axiom 8.5 does not assert the existence of 0^{-1}; so division is not defined when $x = 0$. In the language of Section 5.4, the division function is

$$\text{division} : \mathbf{R} \times (\mathbf{R} - \{0\}) \to \mathbf{R}, \qquad \text{division}(y,x) = y \cdot x^{-1}.$$

Note that $\frac{1}{x} = 1 \cdot x^{-1} = x^{-1}$, and so we usually write x^{-1} as $\frac{1}{x}$.

Project 8.25. Think about why division by 0 ought not to be defined. Come up with an argument that will convince a friend.

8.2 Positive Real Numbers and Ordering

Axiom 8.26. *There exists a subset* $\mathbf{R}_{>0} \subseteq \mathbf{R}$ *satisfying:*

(i) *If* $x, y \in \mathbf{R}_{>0}$ *then* $x + y \in \mathbf{R}_{>0}$.

(ii) *If* $x, y \in \mathbf{R}_{>0}$ *then* $xy \in \mathbf{R}_{>0}$.

(iii) $0 \notin \mathbf{R}_{>0}$.

(iv) *For every* $x \in \mathbf{R}$, *we have* $x \in \mathbf{R}_{>0}$ *or* $x = 0$ *or* $-x \in \mathbf{R}_{>0}$.

The members of $\mathbf{R}_{>0}$ are called **positive real numbers**. A **negative real number** is a real number that is neither positive nor zero.

Proposition 8.27. *For* $x \in \mathbf{R}$, *one and only one of the following is true:* $x \in \mathbf{R}_{>0}$, $-x \in \mathbf{R}_{>0}$, $x = 0$.

Proposition 8.28. $1 \in \mathbf{R}_{>0}$.

By analogy with the definition of "less than" in $\mathbf{Z}$, we write $x < y$ (x **is less than** y) or $y > x$ (y **is greater than** x) if $y - x \in \mathbf{R}_{>0}$, and we write $x \leq y$ (x **is less than or equal to** y) or $y \geq x$ (y **is greater than or equal to** x) if we also allow $x = y$. The analogy between the $<$ relation on $\mathbf{R}$ and $<$ as previously defined on $\mathbf{Z}$ continues:

Proposition 8.29. *Let* $x, y, z \in \mathbf{R}$. *If* $x < y$ *and* $y < z$ *then* $x < z$.

Proposition 8.30. *For each* $x \in \mathbf{R}$ *there exists* $y \in \mathbf{R}$ *such that* $y > x$.

Proposition 8.31. *Let* $x, y \in \mathbf{R}$. *If* $x \leq y \leq x$ *then* $x = y$.

Proposition 8.32. *For all* $x, y, z, w \in \mathbf{R}$:

(i) *If* $x < y$ *then* $x + z < y + z$.

(ii) *If* $x < y$ *and* $z < w$ *then* $x + z < y + w$.

(iii) *If* $0 < x < y$ *and* $0 < z \leq w$ *then* $xz < yw$.

(iv) *If* $x < y$ *and* $z < 0$ *then* $yz < xz$.

Proposition 8.33. *For each* $x, y \in \mathbf{R}$, *exactly one of the following is true:* $x < y$, $x = y$, $x > y$.

Proposition 8.34. *Let* $x \in \mathbf{R}$. *If* $x \neq 0$ *then* $x^2 \in \mathbf{R}_{>0}$.

Proposition 8.35. *The equation* $x^2 = -1$ *has no solution in* $\mathbf{R}$.

Proposition 8.36. *Let* $x, z \in \mathbf{R}_{>0}$, $y \in \mathbf{R}$. *If* $xy = z$, *then* $y \in \mathbf{R}_{>0}$.

Proposition 8.37. *For all* $x, y, z \in \mathbf{R}$:

(i) $-x < -y$ *if and only if* $x > y$.

(ii) *If* $x > 0$ *and* $xy < xz$ *then* $y < z$.

(iii) *If* $x < 0$ *and* $xy < xz$ *then* $z < y$.

(iv) *If* $x \le y$ *and* $0 \le z$ *then* $xz \le yz$.

Proposition 8.38. $\mathbf{R}_{>0} = \{x \in \mathbf{R} : x > 0\}$.

Proposition 8.39. *If* $x \in \mathbf{R}_{>0}$ *then* $x+1 \in \mathbf{R}_{>0}$.

Proposition 8.40.

(i) $x \in \mathbf{R}_{>0}$ *if and only if* $\frac{1}{x} \in \mathbf{R}_{>0}$.

(ii) *Let* $x, y \in \mathbf{R}_{>0}$. *If* $x < y$ *then* $0 < \frac{1}{y} < \frac{1}{x}$.

After proving this proposition, draw graphs of $y = x^2$ *and* $y = x^3$.

Proposition 8.41. *Let* $x \in \mathbf{R}$. *Then* $x^2 < x^3$ *if and only if* $x > 1$.

8.3 Similarities and Differences

A mathematical system that satisfies Axioms 1.1–1.4 is called a commutative ring.

If you compare Axioms 1.1–1.4 (for **Z**) with Axioms 8.1–8.4 (for **R**) you will see that they are identical. They are concerned with addition, subtraction, 0, and 1. It follows that any proposition for **Z** that depends only on Axioms 1.1–1.4 is automatically also true for **R**. In fact, the same holds for $\mathbf{Z}_n$, by Proposition 6.26.

In the same way, Axiom 2.1 and Axiom 8.26 are identical: they concern the positive numbers and ordering. Thus once again we can get "free" theorems for real numbers based on proofs originally given for integers.

Now compare Axiom 1.5 (cancellation) with Axiom 8.5 (multiplicative inverse). As we showed in Proposition 8.7, Axiom 8.5 implies Axiom 1.5. The converse implication is false: for example, the integer 2 does not have a multiplicative inverse in **Z**.

Another notable difference between **Z** and **R** involves the existence of a smallest positive element. By Proposition 2.20, the integer 1 is the smallest positive integer. There is no comparable statement for **R**:

Here we use the same definition for "smallest element" that we used in Section 2.4.

Theorem 8.42. $\mathbf{R}_{>0}$ *does not have a smallest element.*

Proof. Define the real number $2 := 1+1$; by Proposition 8.28, $2 \in \mathbf{R}_{>0}$. Proposition 8.40 implies that $2^{-1} = \frac{1}{2}$ is also positive.

We claim further that $\frac{1}{2} < 1$; otherwise, Proposition 8.32(ii) (with $0 < 1 < 2$ and $0 < 1 \le \frac{1}{2}$) would imply that $1 < 1$, a contradiction.

Thus we have established $0 < \frac{1}{2} < 1$ and can start the actual proof of Theorem 8.42. We will prove it by contradiction. Assume that there exists a smallest element

$s \in \mathbf{R}_{>0}$. Then we can use Proposition 8.32(ii) (with $0 < \frac{1}{2} < 1$ and $0 < s \leq s$) to deduce

$$\frac{1}{2} \cdot s < s .$$

However, $\frac{1}{2} \cdot s \in \mathbf{R}_{>0}$ (by Axiom 8.26(ii)), which contradicts the fact that s is the smallest element in $\mathbf{R}_{>0}$. □

We labeled Theorem 8.42 as a theorem rather than a proposition to emphasize its importance. In many of your advanced mathematics courses—courses with words like *analysis* and *topology* in their titles—the instructor will use Theorem 8.42 regularly. It may not be mentioned explicitly, but it will be used in ε–δ arguments. We will discuss this in more detail in Chapter 10.

Theorem 8.43. *Let $x, y \in \mathbf{R}$ such that $x < y$. There exists $z \in \mathbf{R}$ such that $x < z < y$.*

This theorem implies that the real numbers are "all over the place" in the sense that no matter how close two real numbers are, there are infinitely many real numbers between these two. (See Section 13.1 for the meaning of "infinitely many.")

The analogous statement for $\mathbf{Z}$ is false—this is the content of Corollary 2.22.

The remaining axiom for $\mathbf{Z}$, Axiom 2.15, is concerned with induction; it has no analogue for the real numbers:

Project 8.44. Construct a subset $A \subseteq \mathbf{R}$ that satisfies

(i) $1 \in A$ and

(ii) if $n \in A$ then $n + 1 \in A$,

yet for which $\mathbf{R}_{>0}$ is not a subset of A.

In the next section, we will introduce one more axiom for $\mathbf{R}$, called the Completeness Axiom; it has no useful analogue for $\mathbf{Z}$.

8.4 Upper Bounds

To state our last axiom for $\mathbf{R}$, we need some definitions. Let A be a nonempty subset of $\mathbf{R}$.

(i) The set A is **bounded above** if there exists $b \in \mathbf{R}$ such that for all $a \in A$, $a \leq b$. Any such number b is called an **upper bound** for A.

(ii) The set A is **bounded below** if there exists $b \in \mathbf{R}$ such that for all $a \in A$, $b \leq a$. Any such number b is called a **lower bound** for A.

(iii) The set A is **bounded** if it is both bounded above and bounded below.

(iv) A **least upper bound** for A is a an upper bound that is less than or equal to every upper bound for A.

Least upper bounds are unique if they exist:

Proposition 8.45. *If x_1 and x_2 are least upper bounds for A, then* $x_1 = x_2$.

$\sup(A)$ is often written as $\sup A$, as in Example 8.46. An alternative notation for $\sup(A)$ is $\operatorname{lub}(A)$.

The least upper bound of A is denoted by $\sup(A)$, an abbreviation for **supremum**.

Example 8.46. $\sup\{x \in \mathbf{R} : x < 0\} = 0$.

The least upper bound of a set might not exist. For example:

Proposition 8.47. $\mathbf{R}_{>0}$ *has no upper bound.*

Example 8.48. Consider the sets

$$\{x \in \mathbf{R} : 0 \leq x \leq 1\}$$

and

$$\{x \in \mathbf{R} : 0 \leq x < 1\}.$$

In both cases, the least upper bound is 1. In the first set, the least upper bound lies in the set, while in the second set the least upper bound lies outside. The important fact, illustrated by this example, is that $\sup(A)$ sometimes lies in A but not always. We will say more in the next proposition.

Propositions 8.45 and 8.49 imply that $\max(A)$ is unique if it exists.

A real number $b \in A$ is the **maximum** or **largest** element of A if for all $a \in A$, $a \leq b$. In this case we write $b = \max(A)$.

Proposition 8.49. *Let $A \subseteq \mathbf{R}$ be nonempty. If $\sup(A) \in A$ then $\sup(A)$ is the largest element of A, i.e., $\sup(A) = \max(A)$. Conversely, if A has a largest element then $\max(A) = \sup(A)$ and $\sup(A) \in A$.*

Proposition 8.50. *If the sets A and B are bounded above and $A \subseteq B$, then* $\sup(A) \leq \sup(B)$.

At this point it is useful to define **intervals**. They come in nine types: Let $x < y$. Then

$[x,y]$ is an example of a closed interval; (x,y) is an open interval; and $(x,y]$ is half open. Do not confuse the open interval notation with the coordinate description of a point in the plane.

$$\begin{aligned}
[x,y] &:= \{z \in \mathbf{R} : x \leq z \leq y\} \\
(x,y] &:= \{z \in \mathbf{R} : x < z \leq y\} \\
[x,y) &:= \{z \in \mathbf{R} : x \leq z < y\} \\
(x,y) &:= \{z \in \mathbf{R} : x < z < y\} \\
(-\infty,x] &:= \{z \in \mathbf{R} : z \leq x\} \\
(-\infty,x) &:= \{z \in \mathbf{R} : z < x\} \\
[x,\infty) &:= \{z \in \mathbf{R} : x \leq z\} \\
(x,\infty) &:= \{z \in \mathbf{R} : x < z\} \\
(-\infty,\infty) &:= \mathbf{R}.
\end{aligned}$$

Project 8.51. For a nonempty set $B \subseteq \mathbf{R}$ one can define the **greatest lower bound** $\inf(B)$ (for **infimum**) of B. Give the precise definition for $\inf(B)$ and prove that it is unique if it exists. Also define $\min(B)$ and prove the analogue of Proposition 8.49 for greatest lower bounds and minima.

An alternative notation for $\inf(B)$ *is* $\text{glb}(B)$.

Here is the final axiom for the real numbers.

Axiom 8.52 (Completeness Axiom). *Every nonempty subset of* $\mathbf{R}$ *that is bounded above has a least upper bound.*

This axiom, which concludes our definition of $\mathbf{R}$, is stated here only because those referring back later might forget to include it in the list. It needs discussion, indeed a chapter of its own—Chapter 10.

Proposition 8.53. *Every nonempty subset of* $\mathbf{R}$ *that is bounded below has a greatest lower bound.*

Review Questions. Have you looked carefully at how the axioms for the set of real numbers differ from the axioms for the set of integers? Do you understand the difference between the maximum element of a set of real numbers and the least upper bound of that set?

Weekly reminder: Reading mathematics is not like reading novels or history. You need to think slowly about every sentence. Usually, you will need to reread the same material later, often more than one rereading.

This is a short book. Its core material occupies about 140 pages. Yet it takes a semester for most students to master this material. In summary: read line by line, not page by page.

Chapter 9
Embedding Z in R

I believe that numbers and functions of analysis are not the arbitrary result of our minds; I think that they exist outside of us, with the same character of necessity as the things of objective reality, and we meet them or discover them, and study them, as do the physicists, the chemists and the zoologists.
Charles Hermite (1822–1901), quoted in Morris Kline's *Mathematical Thought from Ancient to Modern Times*, Oxford University Press, 1972, p. 1035.

We have now defined two number systems, **Z** and **R**. Intuitively, we think of the integers as a subset of the real numbers; however, nothing in our axioms tells us explicitly that **Z** can be viewed as a subset of **R**. In fact, at the moment we have no axiomatic reason to think that the integers we named 0 and 1 are the same as the real numbers we named 0 and 1.

Just for now, we will be more careful and write $0_{\mathbf{Z}}$ and $1_{\mathbf{Z}}$ for these special members of **Z**, and $0_{\mathbf{R}}$ and $1_{\mathbf{R}}$ for the corresponding special members of **R**. Informally we are accustomed to identifying $0_{\mathbf{Z}}$ with $0_{\mathbf{R}}$ and identifying $1_{\mathbf{Z}}$ with $1_{\mathbf{R}}$. We will justify this here by giving an *embedding* of **Z** into **R**, that is, a function that maps each integer to the corresponding number in **R**.

Before You Get Started. How could such an embedding function of **Z** into **R** be constructed? From what you know about functions, what properties will such a function have?

M. Beck and R. Geoghegan, *The Art of Proof: Basic Training for Deeper Mathematics*, Undergraduate Texts in Mathematics, DOI 10.1007/978-1-4419-7023-7_9,

9.1 Injections and Surjections

An injective function is also called an injection, *an* embedding, *or a* one-to-one function.

A function $f : A \to B$ is **injective** if

$$\text{for all } a_1, a_2 \in A, \ a_1 \neq a_2 \text{ implies } f(a_1) \neq f(a_2) .$$

It is equivalent to require here the contrapositive condition (see Section 3.2), namely, a function $f : A \to B$ is injective if

$$\text{for all } a_1, a_2 \in A, \ f(a_1) = f(a_2) \text{ implies } a_1 = a_2 .$$

Example 9.1. The function

$$f : \mathbf{Z} \to \mathbf{Z} \qquad \text{defined by} \qquad f(n) = 3n$$

is injective. The function

$$g : \mathbf{Z} \to \mathbf{Z} \qquad \text{defined by} \qquad g(n) = n^2$$

is not injective. (Prove both statements.)

A surjective function is also called a surjection *or an* onto function.

A function $f : A \to B$ is **surjective** if

$$\text{for each } b \in B \text{ there exists } a \in A \text{ such that } f(a) = b .$$

The image of a function is also called its range.

The **image** of a function $f : A \to B$, denoted by $f(A)$, is the set

$$f(A) := \{ f(a) : a \in A \} .$$

A bijective function is also called a bijection *or a* one-to-one correspondence.

Thus $f : A \to B$ is surjective if and only if $f(A) = B$, or, in words, if the image of f equals its codomain. The function $f : A \to B$ is **bijective** if f is both injective and surjective.

Example 9.2. A seemingly trivial but important function is the **identity function** on a set A, namely the function

$$\mathrm{id}_A : A \to A \qquad \text{defined by} \qquad \mathrm{id}_A(a) = a \ \text{ for all } a \in A.$$

This function is bijective.

Project 9.3. Determine which of the following functions are injective, surjective, or bijective. Justify your assertions.

(i) $f : \mathbf{Z} \to \mathbf{Z}, \ f(n) = n^2$.

(ii) $f : \mathbf{Z} \to \mathbf{Z}_{\geq 0}, \ f(n) = n^2$.

(iii) $f : \mathbf{Z}_{\geq 0} \to \mathbf{Z}_{\geq 0}$, $f(n) = n^2$.

(iv) $f : \mathbf{R} \to \mathbf{R}$, $f(x) = 3x + 1$.

(v) $f : \mathbf{R}_{\geq 0} \to \mathbf{R}$, $f(x) = 3x + 1$.

(vi) $f : \mathbf{Z} \to \mathbf{Z}$, $f(x) = 3x + 1$.

Project 9.4. Construct (many) functions that are

(i) bijective;

(ii) injective, but not surjective;

(iii) surjective, but not injective;

(iv) neither injective nor surjective.

Justify your claims.

Example 9.5. Let $f : \mathbf{R} \to \mathbf{R}$ be a function. Consider the graph of f; it is a subset of the Cartesian product $\mathbf{R} \times \mathbf{R}$, which may be identified with the plane. The function f is injective if and only if no horizontal line crosses the graph in two or more places. The function f is surjective if and only if every horizontal line crosses the graph.

Project 9.6. Which differentiable functions $\mathbf{R} \to \mathbf{R}$ are bijections?

$f : \mathbf{R} \to \mathbf{R}$ is differentiable if its derivative $f'(x)$ exists for every $x \in \mathbf{R}$.

The **composition** of two functions $f : A \to B$ and $g : B \to C$ is the function

$$g \circ f : A \to C \quad \text{defined by} \quad (g \circ f)(a) = g(f(a)) \text{ for all } a \in A .$$

Proposition 9.7.

(i) *If $f : A \to B$ is injective and $g : B \to C$ is injective then $g \circ f : A \to C$ is injective.*

(ii) *If $f : A \to B$ is surjective and $g : B \to C$ is surjective then $g \circ f : A \to C$ is surjective.*

(iii) *If $f : A \to B$ is bijective and $g : B \to C$ is bijective then $g \circ f : A \to C$ is bijective.*

Proof of (i). Assume that $f : A \to B$ and $g : B \to C$ are injective functions. Recall what this means: for each $a_1, a_2 \in A$,

$$f(a_1) = f(a_2) \qquad \text{implies} \qquad a_1 = a_2 , \tag{9.1}$$

and for each $b_1, b_2 \in B$,

$$g(b_1) = g(b_2) \qquad \text{implies} \qquad b_1 = b_2 . \tag{9.2}$$

To prove that $g \circ f$ is injective, we assume $(g \circ f)(a_1) = (g \circ f)(a_2)$ and we will show that $a_1 = a_2$. So assume that $g(f(a_1)) = g(f(a_2))$. Then by applying (9.2) to $b_1 = f(a_1)$ and $b_2 = f(a_2)$, we conclude that

$$b_1 = b_2, \qquad \text{that is,} \qquad f(a_1) = f(a_2).$$

But now by (9.1), $a_1 = a_2$. □

A **left inverse** of a function $f : A \to B$ is a function $g : B \to A$ such that

$$g \circ f = \mathrm{id}_A .$$

A **right inverse** of a function $f : A \to B$ is a function $g : B \to A$ such that

$$f \circ g = \mathrm{id}_B .$$

A **(two-sided) inverse** of f is a function that is both a left inverse and a right inverse of f.

This example of a left inverse g can be modified, so there are, in fact, infinitely many left inverses of f.

Example 9.8. Let $f : \mathbf{Z} \to \mathbf{Z}$ be defined by $f(n) = 2n$. Then the function $g : \mathbf{Z} \to \mathbf{Z}$ given by

$$g(n) := \begin{cases} \frac{m}{2} & \text{if } m \text{ is even,} \\ 34 & \text{if } m \text{ is odd} \end{cases}$$

is a left inverse of f, because for all $n \in \mathbf{Z}$

$$(g \circ f)(n) = g(f(n)) = g(2n) = \frac{2n}{2} = n .$$

Note that g is *not* a right inverse of f. In fact, Proposition 9.10 below implies that f does not have a right inverse.

Project 9.9. Find left/right inverses (if they exist) for each of your examples in Project 9.4.

Proposition 9.10.

(i) *f is injective if and only if f has a left inverse.*

(ii) *f is surjective if and only if f has a right inverse.*

(iii) *f is bijective if and only if f has an inverse.*

Proof. (i) Let $f : A \to B$ be injective. Fix an $a_0 \in A$ and define the function $g : B \to A$ by

$$g(b) := \begin{cases} a & \text{if } b \text{ is in the image of } f \text{ and } f(a) = b, \\ a_0 & \text{otherwise.} \end{cases}$$

Then g is a well-defined function, because f is injective, and by construction we have $g \circ f = \mathrm{id}_A$.

Conversely, assume that $f : A \to B$ has a left inverse g, that is, $g : B \to A$ is a function such that $g \circ f = \mathrm{id}_A$. Assume that $a_1, a_2 \in A$ satisfy $f(a_1) = f(a_2)$; we will show that this equation implies $a_1 = a_2$, and this will prove that f is injective. Because g is a function, $f(a_1) = f(a_2)$ implies that

$$(g \circ f)(a_1) = g(f(a_1)) = g(f(a_2)) = (g \circ f)(a_2).$$

Comparing the left-hand side of this equation with the right-hand side yields $a_1 = a_2$, since $g \circ f = \mathrm{id}_A$.

(ii) Let $f : A \to B$ be surjective. We will construct a function $g : B \to A$ as follows: Given $b \in B$, choose an $a \in A$ such that $f(a) = b$ (we can possibly find more than one such a, in which case we choose one). We define this a to be the image of b under the function g, that is, we define $g(b) = a$. With this definition, we deduce

The Axiom of Choice *is the assertion that this way of defining the function g is legitimate. Lurking here is a deep issue in set theory.*

$$(f \circ g)(b) = f(g(b)) = f(a) = b,$$

and so g is a right inverse of f.

Conversely, assume that $f : A \to B$ has a right inverse g, that is, $g : B \to A$ is a function such that $f \circ g = \mathrm{id}_B$. We need to show that f is surjective, that is, given $b \in B$, we need to find $a \in A$ such that $f(a) = b$. Given such a $b \in B$, we define $a = g(b)$. Then by construction

$$f(a) = f(g(b)) = (f \circ g)(b) = b.$$

(iii) Let $f : A \to B$ be bijective. Then our constructions of g described in (i) and (ii) coincide: namely, we define a function $g : B \to A$ by

$$g(b) = a \qquad \text{if and only if} \qquad f(a) = b.$$

This function g is well defined because

- f is surjective (and so we can define $g(b)$ for every b) and
- f is injective (and so for every b there exists a unique a such that $g(b) = a$).

Thus our construction of a left inverse in (i) and of a right inverse in (ii) yield the same function, that is, f has an inverse.

Conversely, assume that $f : A \to B$ has an inverse g. Then it follows from (i) that f is injective and from (ii) that f is surjective. Thus f is bijective. □

Proposition 9.11. *If a function is bijective then its inverse is unique.*

Proposition 9.12. *Let A and B be sets. There exists an injection from A to B if and only if there exists a surjection from B to A.*

Project 9.13. Let $f : A \to B$ and $g : B \to C$. Decide whether each of the following is true or false; in each case prove the statement or give a counterexample.

(i) If f is injective and g is surjective then $g \circ f$ is surjective.

(ii) If $g \circ f$ is bijective then g is surjective and f is injective.

9.2 The Relationship between Z and R

We want to embed **Z** into **R**. To do this, we define an injective function $e : \mathbf{Z} \to \mathbf{R}$ as follows:

Here the first addition takes place in **Z**, *while the second addition happens in* **R**.

(i) Define e on $\mathbf{Z}_{\geq 0}$ recursively: $e(0_{\mathbf{Z}}) := 0_{\mathbf{R}}$ and, assuming $e(n)$ defined for a fixed $n \in \mathbf{Z}_{\geq 0}$, define

$$e(n + 1_{\mathbf{Z}}) := e(n) + 1_{\mathbf{R}}.$$

(ii) If $k \in \mathbf{Z}$ and $k < 0$, define $e(k) := -e(-k)$.

Proposition 9.14.

(i) $e(1_{\mathbf{Z}}) = 1_{\mathbf{R}}$.

(ii) $e(-1_{\mathbf{Z}}) = -1_{\mathbf{R}}$.

Proposition 9.15. *If $k \in \mathbf{N}$ then $e(k) \in \mathbf{R}_{>0}$.*

Proposition 9.16. *For all $k \in \mathbf{Z}$,*

(i) $e(k + 1_{\mathbf{Z}}) = e(k) + 1_{\mathbf{R}}$.

(ii) $e(k - 1_{\mathbf{Z}}) = e(k) - 1_{\mathbf{R}}$.

(iii) $e(k) = -e(-k)$.

The point of this proposition is that the statements occurring in the definition of e hold for *all* $k \in \mathbf{Z}$.

Proof. (i) The first equation holds by definition when $k \in \mathbf{Z}_{\geq 0}$. If $k = -1$ then

$$e(k+1) = e(0) = 0 = -1 + 1 = e(k) + 1.$$

If $k < -1$ then $k + 1$ is negative and

$$\begin{aligned} e(k+1) &= -e(-(k+1)) \\ &= -e(-k-1) \\ &= -(e(-k) - 1) \\ &= -e(-k) + 1 \\ &= e(k) + 1. \end{aligned} \tag{9.3}$$

Here (9.3) follows from $e(-k) = e((-k-1)+1) = e(-k-1)+1$ (note that $-k-1 > 0$).

(ii) The second equation follows from part (i) with

$$e(k) = e((k-1)+1) = e(k-1)+1\,.$$

(iii) The equation $e(k) = -e(-k)$ holds by definition for negative k. We will prove it now for $k \geq 0$ by induction. The base case $k = 0$ follows because $0 = -0$ (in $\mathbf{Z}$ as well as in $\mathbf{R}$), and so $e(0) = -e(-0)$.

For the induction step, assume that $e(n) = -e(-n)$. Then, by applying parts (i) and (ii),

$$\begin{aligned} e(n+1) &= e(n)+1 = -e(-n)+1 = -\left(e(-n)-1\right) = -e(-n-1) \\ &= -e\left(-(n+1)\right). \end{aligned}$$

□

Proposition 9.17. *The function e preserves addition: for all* $m, k \in \mathbf{Z}$,

$$e\,(m+k) = e(m) + e(k)\,,$$

where $+$ *on the left-hand side refers to addition in* $\mathbf{Z}$*, and* $+$ *on the right-hand side refers to addition in* $\mathbf{R}$.

Proof. Fix $m \in \mathbf{Z}$. We will prove that for all $k \in \mathbf{Z}$, $e\,(m+k) = e(m) + e(k)$. First, the proposition holds for $k = 0$, since $e(0) = 0$.

Next, we prove

$$P(k) : e\,(m+k) = e(m) + e(k)$$

by induction on $k \in \mathbf{N}$, which will establish the proposition for *positive* k. The base case $P(1)$ follows by definition.

For the induction step, assume $P(n)$. Then, by Proposition 9.16(i),

$$\begin{aligned} e\,(m+(n+1)) &= e\,(m+n)+1 \\ &= e(m)+e(n)+1 \qquad (9.4) \\ &= e(m)+e(n+1)\,. \end{aligned}$$

Here (9.4) follows from the induction hypothesis.

Finally, we prove

$$Q(k) : e\,(m+(-k)) = e(m) + e(-k)$$

by induction on $k \in \mathbf{N}$, which will establish the proposition. The base case $Q(1)$ follows from Proposition 9.16(ii). For the induction step, assume $Q(n)$. Then, again by Proposition 9.16(ii),

$$\begin{aligned} e\left(m+\left(-(n+1)\right)\right) &= e\left(m+(-n)-1\right) \\ &= e\left(m+(-n)\right)-1 \\ &= e(m)+e(-n)-1 \qquad (9.5) \\ &= e(m)+e(-n-1) \\ &= e(m)+e\left(-(n+1)\right), \end{aligned}$$

where (9.5) follows from the induction hypothesis. □

Proposition 9.18. *The function e preserves multiplication: for all* $m,k \in \mathbf{Z}$,

$$e(m \cdot k) = e(m) \cdot e(k),$$

where $\cdot$ *on the left-hand side refers to multiplication in* **Z**, *whereas* $\cdot$ *on the right-hand side refers to multiplication in* **R**.

Proposition 9.19. *The function e preserves order:* $m,k \in \mathbf{Z}$ *satisfy*

$$m < k \qquad \text{if and only if} \qquad e(m) < e(k).$$

Here $<$ *on the left-hand side refers to the less-than relation in* **Z**, *whereas* $<$ *on the right-hand side refers to the less-than relation in* **R**.

Proof of the "only if" statement. Let $m < k$. Then $e(k-m) > 0$ (by Proposition 9.15), which we can rewrite as

$$e(k-m) \stackrel{\text{Prop. 9.17}}{=} e(k)+e(-m) \stackrel{\text{Prop. 9.16}(iii)}{=} e(k)-e(m) > 0,$$

that is, $e(m) < e(k)$. □

Corollary 9.20. *The function e is injective.*

Thus **R** has a subset $e(\mathbf{Z})$ that behaves exactly like **Z** with respect to addition, multiplication, and order. It follows that $e(\mathbf{Z})$ behaves like **Z** with respect to every property of **Z** we have discussed in this book.

9.3 Apples and Oranges Are All Just Fruit

In this book we have studied **Z** and **R** separately as if they were apples and oranges. Now, in this chapter, we have embedded **Z** in **R**. Usually, people do not think this way. They simply think of **Z** as a subset of **R** (as you have always done). We will do that from now on, and so we will not distinguish between $n \in \mathbf{Z}$ and $e(n) \in \mathbf{R}$.

We drop notational distinctions such as $0_{\mathbf{Z}}$ and $1_{\mathbf{R}}$ and write $\mathbf{Z} \subseteq \mathbf{R}$. The apples and oranges have become generic fruit.

Review Question. Have you understood how $\mathbf{Z}$ becomes a subset of $\mathbf{R}$?

Weekly reminder: Reading mathematics is not like reading novels or history. You need to think slowly about every sentence. Usually, you will need to reread the same material later, often more than one rereading.

This is a short book. Its core material occupies about 140 pages. Yet it takes a semester for most students to master this material. In summary: read line by line, not page by page.

Chapter 10

Limits and Other Consequences of Completeness

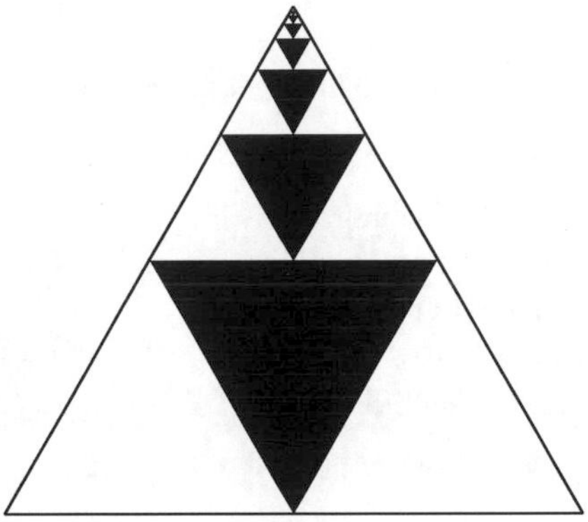

Before You Get Started. You probably know that the sequence of real numbers $\frac{n-1}{n}$ converges to the real number 1 as n gets larger. What exactly does this mean? Can you guess a definition of "converges" that does not use words like "nearer and nearer to" or "approaches"?

Think about the picture above. Each black triangle occupies $\frac{1}{4}$ of the area of its "parent" triangle. If the area of the biggest triangle is 1, then the sum of the areas of all the black triangles should be thought of as $\frac{1}{4} + \frac{1}{4^2} + \frac{1}{4^3} + \cdots$. You might remember from previous math courses that the sum of this series is $\frac{1}{3}$. Can you see in the picture that the blackened area constitutes $\frac{1}{3}$ of the total area of the biggest white triangle? (Look at the white-black-white horizontal rows of triangles.) This is an example of convergence.

M. Beck and R. Geoghegan, *The Art of Proof: Basic Training for Deeper Mathematics*, Undergraduate Texts in Mathematics, DOI 10.1007/978-1-4419-7023-7_10,

10.1 The Integers Are Unbounded

This chapter is devoted to consequences of the Completeness Axiom 8.52: *Every nonempty subset of* **R** *that is bounded above has a least upper bound.*

Theorem 10.1. **N**, *considered as a subset of* **R**, *is not bounded above.*

Proof. We will prove this by contradiction. Suppose **N** were bounded above. Then, by Axiom 8.52, **N** would have a least upper bound: call it u. The interval

$$\left(u-\tfrac{1}{2},u\right] = \left\{x \in \mathbf{R} : u-\tfrac{1}{2} < x \le u\right\}$$

must contain some $n \in \mathbf{N}$ since otherwise $u-\frac{1}{2}$ would be an upper bound for **N**, contradicting the fact that u is the least upper bound. But if $u-\frac{1}{2} < n$ then $u+\frac{1}{2} < n+1$, so $u < n+1$ (because $\frac{1}{2} > 0$ by Corollary 8.40). By Axiom 2.1 and Proposition 2.3, $n+1 \in \mathbf{N}$, so u is not an upper bound for **N**, which is a contradiction. □

The fact that for each $x \in \mathbf{R}$, there exists an integer greater than x is called the Archimedean property *of* **R**.

Take a moment to think about Theorem 10.1. In Proposition 2.5 we saw that there is no largest natural number. If the real numbers are pictured by a horizontal line (draw one) and if the first few natural numbers are marked on that line (do it: $1,2,3,4,\dots$), we have to rule out the possibility that there is a real number larger than all of them, and that is precisely the statement of Theorem 10.1.

Since $\mathbf{N} \subseteq \mathbf{Z}$ we deduce the following corollary.

Corollary 10.2. **Z** *is not bounded above.*

We proved earlier that 1 is the least element of **N**. Note that **Z** does not have a least element. Even more is true:

Corollary 10.3. **Z** *is not bounded below.*

Another consequence of the unboundedness of **N** is the following useful proposition.

Proposition 10.4. *For each $\varepsilon > 0$, there exists $n \in \mathbf{N}$ such that $\frac{1}{n} < \varepsilon$.*

Proof. Let $\varepsilon > 0$. By Proposition 10.1, there exists $n \in \mathbf{N}$ such that $n > \frac{1}{\varepsilon}$, or equivalently by Proposition 8.32(ii), $\frac{1}{n} < \varepsilon$. □

10.2 Absolute Value

This definition is analogous to our definition of $|x|$ when $x \in \mathbf{Z}$, which we gave in Section 6.1.

The **absolute value** of $x \in \mathbf{R}$, denoted by $|x|$, is defined to be x if $x \ge 0$ and to be

$-x$ if $x < 0$. This definition implies that $|x| \geq 0$ always, because the negative of a negative number is positive.

Proposition 10.5. *Let $x, y \in \mathbf{R}_{\geq 0}$. Then $x < y$ if and only if $x^2 < y^2$.*

Proposition 10.6. *For all $x \in \mathbf{R}$, $|x|^2 = x^2$.*

Proposition 10.7. *Let $x, y \in \mathbf{R}$. Then $|x| < |y|$ if and only if $x^2 < y^2$.*

Proposition 10.8. *For all $x, y \in \mathbf{R}$:*

(i) $|x| = 0$ *if and only if* $x = 0$.

(ii) $|xy| = |x|\,|y|$.

(iii) $-|x| \leq x \leq |x|$.

(iv) $|x + y| \leq |x| + |y|$.

(v) *If* $-y < x < y$ *then* $|x| < |y|$.

Part (iv) is called the triangle inequality*; this name makes more literal sense when we allow x and y to be complex numbers, as we will see in Proposition C.11.*

Proof of parts (i) *and* (iv). (i) If $x = 0$ then, by definition, $|x| = x = 0$.

Conversely, if $x \neq 0$ then either $x > 0$, in which case $|x| = x > 0$, or $x < 0$, in which case $|x| = -x > 0$ by Proposition 8.32(iii). In both cases we conclude that $|x| > 0$, so in particular, $|x| \neq 0$.

(iv) By Proposition 10.6,

$$|x+y|^2 = (x+y)^2 = x^2 + 2xy + y^2 = |x|^2 + 2xy + |y|^2.$$

By part (iii), $2xy < |2xy| = 2|x||y|$; here the last equality follows from part (ii). Hence

$$|x+y|^2 = |x|^2 + 2xy + |y|^2 \leq |x|^2 + 2|x||y| + |y|^2 = (|x| + |y|)^2.$$

Proposition 10.5 then implies $|x + y| \leq |x| + |y|$. □

Proposition 10.9. *Let $x \in \mathbf{R}$ be such that $0 \leq x \leq 1$, and let $m, n \in \mathbf{N}$ be such that $m \geq n$. Then $x^m \leq x^n$.*

10.3 Distance

Proposition 10.10. *Let $x, y, z \in \mathbf{R}$.*

(i) $|x - y| = 0$ *if and only if* $x = y$.

(ii) $|x-y| = |y-x|$.

(iii) $|x-z| \le |x-y| + |y-z|$.

(iv) $|x-y| \ge \big||x|-|y|\big|$.

One of the beautiful things about mathematics is that it involves both algebra and geometry; in fact, there are times when one wants to express a geometrical statement using the language of algebra, and there are other times when one wants to express an algebraic statement in the language of geometry. Absolute value provides a good example. In mathematics we think of $|x-y|$ as the **distance** from the point x to the point y on the line $\mathbf{R}$. In fact, let us make this the definition of the word "distance." This agrees with the everyday definition of that word: distance is never negative (try going for a walk -2 miles in length), and the distance from a point to a different point is never 0. In this language, we can reformulate Proposition 10.10:

Try putting Proposition 10.10(iv) in words: sometimes algebraic language is easier.

(i) The distance from x to y is 0 if and only if x equals y.

(ii) The distance from x to y equals the distance from y to x.

(iii) The distance from x to z is at most the sum of the distances from x to y and from y to z.

Proposition 10.11. *Let $x, y \in \mathbf{R}$. Then $x = y$ if and only if for every $\varepsilon > 0$ we have $|x-y| < \varepsilon$.*

Project 10.12. There is a legend that in the early days of cars a road sign in Ireland read, "It is forbidden to exceed any speed over 30 miles per hour." What was the speed limit? Prove your answer.

10.4 Limits

Let $(x_k)_{k=1}^{\infty}$ be a sequence in $\mathbf{R}$, i.e., a function with domain $\mathbf{N}$ and codomain $\mathbf{R}$. Intuitively $L \in \mathbf{R}$ is the limit of this sequence if the numbers x_k get closer and closer to L as k increases. We will make this precise as follows: We say that (x_k) **converges to** L if

$$\text{for each } \varepsilon > 0 \text{ there exists } N \in \mathbf{N} \text{ such that for each } n \ge N,\ |x_n - L| < \varepsilon,$$

or, in quantifier language,

$$\forall \varepsilon > 0\ \exists N \in \mathbf{N} \text{ such that } \forall n \ge N,\ |x_n - L| < \varepsilon.$$

In the language of geometry, for each $\varepsilon > 0$, no matter how small, there is a natural number N such that whenever $n \ge N$, the distance from x_n to L is less than ε.

When (x_k) converges to L, we call L the **limit** of the sequence (x_k), and we write

$$\lim_{k\to\infty} x_k = L.$$

Our language here suggests that L is unique, which is the content of Proposition 10.14.

Here are some examples of convergent sequences.

Proposition 10.13.

(i) $\displaystyle\lim_{k\to\infty} \frac{1}{k} = 0.$

(ii) $\displaystyle\lim_{k\to\infty} \frac{k-1}{k} = 1.$

(iii) $\displaystyle\lim_{k\to\infty} \left(\frac{1}{4^k}\right) = 0.$

Proof of (i) *and* (iii). (i) Let $\varepsilon > 0$ be given. By Proposition 10.4, there exists an integer $N > \frac{1}{\varepsilon}$, and we have for $n \geq N$,

$$\left|\frac{1}{n} - 0\right| = \frac{1}{n} \leq \frac{1}{N} < \varepsilon. \tag{10.1}$$

(The distance from $\frac{1}{n}$ to 0 is less than ε.)

(iii) Let $\varepsilon > 0$ be given. By Proposition 10.4, there exists an integer $N > \frac{1}{\varepsilon}$, and we have for $n \geq N$,

$$\left|\frac{1}{4^n} - 0\right| = \frac{1}{4^n} < \frac{1}{n} \leq \frac{1}{N} < \varepsilon.$$

(The distance from $\frac{1}{4^n}$ to 0 is less than ε.) Here the inequality $\frac{1}{4^n} < \frac{1}{n}$ follows from Propositions 4.8 and 8.40(ii). □

Reflecting on the proof of (i), in practice it is typical that we work out the steps for the inequality (10.1) *first*—these steps usually lead to the required condition for N in terms of ε. For the proof of (i), the calculation

$$\left|\frac{1}{n} - 0\right| = \frac{1}{n} \leq \frac{1}{N}$$

is natural from the given data; at this point all that remains for us to do is to bound the expression on the right by ε, and this gives the condition $N > \frac{1}{\varepsilon}$ in this example.

Sometimes we are interested only in the *fact* that a sequence converges, rather than what it converges to. So to say that (x_k) **converges** means that there exists $L \in \mathbf{R}$ such that (x_k) converges to L. If no such L exists, we say that the sequence **diverges**. Thus the statement "(x_k) diverges" is the negation of the statement "there exists $L \in \mathbf{R}$ such that (x_k) converges to L," i.e.,

$$\forall L \in \mathbf{R}\ \exists \varepsilon > 0 \text{ such that } \forall N \in \mathbf{N}\ \exists n \geq N \text{ such that } |x_n - L| \geq \varepsilon\,.$$

Proposition 10.14 (Uniqueness of limits). *If (x_k) converges to L and to L' then $L = L'$.*

Project 10.15. Find the limits of your favorite sequences from calculus, such as $\left(\frac{1}{3k^2}\right)_{k=0}^{\infty}$, $\left(3^{2-\frac{1}{k}}\right)_{k=0}^{\infty}$, or $\left(\frac{1}{\sqrt{k}}+7\right)_{k=0}^{\infty}$. Find sequences that diverge. Prove your assertions.

Proposition 10.16. *If the sequence (x_k) converges to L, then $\lim\limits_{k\to\infty} x_{k+1} = L$.*

The next limit we will compute is fundamental; for example, we will need it in Section 12.1 when we discuss infinite geometric series. We will give two proofs, one of which uses the following useful inequality.

Proposition 10.17 (Bernoulli's inequality). *Let $x \in \mathbf{R}_{\geq 0}$ and $k \in \mathbf{Z}_{\geq 0}$. Then*

$$(1+x)^k \geq 1 + kx\,.$$

Proposition 10.18. *If $|x| < 1$ then $\lim\limits_{k\to\infty} x^k = 0$.*

First proof. The case $x = 0$ (is easy and) follows from Proposition 10.23(i) below, so we may assume that $0 < |x| < 1$. Then the following inequality follows from Proposition 10.17 for $N \geq 0$:

$$\left(\frac{1}{|x|}\right)^N = \left(1 + \frac{1-|x|}{|x|}\right)^N \geq 1 + \left(\frac{1-|x|}{|x|}\right)N > \left(\frac{1-|x|}{|x|}\right)N\,. \tag{10.2}$$

We could also use logarithms for this (ε, N) argument, but logarithms have not been defined.

To prove $\lim_{k\to\infty} x^k = 0$, let $\varepsilon > 0$. By Proposition 10.4, there exists an integer

$$N > \frac{|x|}{1-|x|}\frac{1}{\varepsilon}\,. \tag{10.3}$$

Then for $n \geq N$,

$$|x^n - 0| = |x|^n \overset{\text{Prop. 10.9}}{\leq} |x|^N \overset{(10.2)}{<} \frac{|x|}{1-|x|}\frac{1}{N} \overset{(10.3)}{<} \varepsilon\,. \qquad \square$$

We will give a second proof of Proposition 10.18 after we have built up some more machinery for sequences, for example, this famous principle: *a monotonic bounded sequence always converges*. We explain:

The sequence $(x_k)_{k=0}^{\infty}$ is **bounded** if there exist $l, u \in \mathbf{R}$ such that $l \leq x_k \leq u$ for all $k \geq 0$.

The sequence $(x_k)_{k=0}^{\infty}$ is **increasing** if

$$x_{k+1} \geq x_k \qquad \text{for all } k \geq 0,$$

and **decreasing** if

$$x_{k+1} \leq x_k \qquad \text{for all } k \geq 0,$$

A sequence is **monotonic** if it is either increasing or decreasing.

Theorem 10.19. *Every increasing bounded sequence converges.*

Proof. Assume that the sequence $(x_k)_{k=0}^{\infty}$ is increasing and bounded. Because the set

$$A := \{x_k : k \geq 0\} \subseteq \mathbf{R}$$

is bounded, it has a least upper bound s by Axiom 8.52. We claim that $s = \lim_{k\to\infty} x_k$. To prove this, let $\varepsilon > 0$. Then $s - \varepsilon$ is *not* an upper bound for A (since $s = \sup A$), and so there exists N such that $x_N > s - \varepsilon$. But $(x_k)_{k=0}^{\infty}$ is increasing, so $x_n \geq x_N$ for all $n \geq N$. In summary, we have for $n \geq N$,

$$s - \varepsilon < x_N \leq x_n \leq s < s + \varepsilon,$$

and so $|x_n - s| < \varepsilon$, by Proposition 10.8(v). We have proved that given any $\varepsilon > 0$, there exists N such that for $n \geq N$, $|x_n - s| < \varepsilon$, as claimed. □

Project 10.20. Prove the analogous statement for *decreasing* bounded sequences. In summary, we then know that every monotonic bounded sequence converges.

It is important to note that Theorem 10.19 is an *existence theorem*: we proved that the sequence converges without finding its limit. We can do this because Axiom 8.52 asserts the existence of real numbers without providing a method for specifying them.

Proposition 10.21. *Let* $L = \lim_{k\to\infty} x_k$.

(i) *If* $(x_k)_{k=0}^{\infty}$ *is increasing then* $x_k \leq L$ *for all* $k \geq 0$.

(ii) *If* $(x_k)_{k=0}^{\infty}$ *is decreasing then* $x_k \geq L$ *for all* $k \geq 0$.

Proposition 10.22. *If a sequence converges, then it is bounded.*

Proposition 10.23. *Let* $\lim_{k\to\infty} a_k = A$, $\lim_{k\to\infty} b_k = B$, *and let* $c \in \mathbf{R}$ *be fixed.*

(i) $\lim_{k\to\infty} c = c$.

(ii) $\lim_{k\to\infty} (c\,a_k) = c\,A$.

(iii) $\lim_{k\to\infty}(a_k+b_k)=A+B$.

(iv) $\lim_{k\to\infty}(a_k b_k)=AB$.

(v) *If* $A\neq 0$, *then* $\lim_{k\to\infty}\frac{1}{a_k}=\frac{1}{A}$.

Proof of (ii). *Case 1:* $c\neq 0$. We know that $A=\lim_{k\to\infty}a_k$, i.e.,

$$\forall\varepsilon>0\ \exists N\in\mathbf{N}\text{ such that }\forall n\geq N,\ |a_n-A|<\varepsilon. \tag{10.4}$$

We are to prove $\lim_{k\to\infty}(c\,a_k)=cA$. Let $\eta>0$. By (10.4), there exists $N\in\mathbf{N}$ such that for all $n\geq N$,

We are applying (10.4) to the number $\varepsilon=\frac{\eta}{|c|}$, *which we can do because* $c\neq 0$ *and* $\frac{\eta}{|c|}>0$.

$$|a_n-A|<\frac{\eta}{|c|},$$

and so

$$|c\,a_n-cA|=|c|\,|a_n-A|<|c|\frac{\eta}{|c|}=\eta.$$

Since this works for any $\eta>0$, we have proved

$$\forall\eta>0\ \exists N\in\mathbf{N}\text{ such that }\forall n\geq N,\ |c\,a_n-cA|<\eta.$$

Case 2: $c=0$. Then $c\,a_k=0$ for all k, and $cA=0$. □

Proof of (iv). By Proposition 10.22, (b_n) is bounded because it converges. Therefore there exists $M>0$ such that $|b_n|\leq M$ for all $n\in\mathbf{N}$.

Case 1: $A=0$. Given $\varepsilon>0$, we know that there exists $N\in\mathbf{N}$ such that for all $n\geq N$,

$$|a_n|<\frac{\varepsilon}{M}.$$

Then for all $n\geq N$,

$$|a_nb_n-AB|=|a_nb_n|=|a_n|\,|b_n|<\frac{\varepsilon}{M}M=\varepsilon.$$

Case 2: $A\neq 0$. Given $\varepsilon>0$, we know that there exists $N_1\in\mathbf{N}$ such that for all $n\geq N_1$,

$$|a_n-A|<\frac{\varepsilon}{2M},$$

and there exists $N_2\in\mathbf{N}$ such that for all $n\geq N_2$,

$$|b_n-B|<\frac{\varepsilon}{2|A|}.$$

Now let $N=\max(N_1,N_2)$. Then for all $n\geq N$,

$$|a_n b_n - AB| = |a_n b_n - Ab_n + Ab_n - AB| = |(a_n - A)\,b_n + (b_n - B)\,A|$$
$$\leq |a_n - A|\,|b_n| + |b_n - B|\,|A| < \frac{\varepsilon}{2M} M + \frac{\varepsilon}{2|A|} |A| = \varepsilon\,. \qquad \square$$

Proposition 10.23 is useful in limit computations. For example, from what we have already proved, we can conclude that

$$\lim_{k\to\infty} \left(\frac{1}{3k^2} - \frac{1}{k} + 2 \right) = \lim_{k\to\infty} \frac{1}{3k^2} - \lim_{k\to\infty} \frac{1}{k} + \lim_{k\to\infty} 2 = 0 + 0 + 2 = 2\,.$$

Second proof of Proposition 10.18. Let $|x| < 1$. Proposition 10.9 implies

$$|x|^{k+1} \leq |x|^k \qquad \text{for } k \geq 0\,,$$

that is, the sequence $\left(|x|^k\right)_{k=0}^{\infty}$ is decreasing. Furthermore, $|x|^k < 1$, and so by Project 10.20, $\left(|x|^k\right)_{k=0}^{\infty}$ converges. Let

$$L := \lim_{k\to\infty} |x|^k\,;$$

Can you see how $\lim_{k\to\infty} |x|^k = 0$ implies $\lim_{k\to\infty} x^k = 0$?

our goal is to prove $L = 0$. By Propositions 10.16 and 10.23(ii),

$$L = \lim_{k\to\infty} |x|^{k+1} = \lim_{k\to\infty} |x|\,|x|^k = |x|\,L\,.$$

This implies that $L(1 - |x|) = 0$ and thus, since $|x| \neq 1$, $L = 0$. $\square$

Project 10.24. In calculus you learned about sequences (x_k) that "blow up" in the sense that $\lim_{k\to\infty} x_k = \infty$; an example is the sequence given by $x_k = k^2$. We think of this sequence as "converging to infinity"; in this sense people like to say that the limit $\lim_{k\to\infty} x_k$ exists (as opposed to, for example, $\lim_{k\to\infty} (-1)^k k^2$). Come up with a solid mathematical definition for $\lim_{k\to\infty} x_k = \infty$ and prove that $\lim_{k\to\infty} k^2 = \infty$.

10.5 Square Roots

As a consequence of Axiom 8.52 we prove the existence of square roots. We first define the **square root** of 2 by

$$\sqrt{2} := \sup\left\{ x \in \mathbf{R} : x^2 < 2 \right\}.$$

Theorem 10.25. *The real number $\sqrt{2}$ is well defined, positive, and $\sqrt{2}^2 = 2$.*

Note that $-\sqrt{2}$ is also a solution to the equation $x^2 = 2$, but $\sqrt{2}$ means the positive solution.

The symbol 3 *means* $+3$ *in the same way that the symbol $\sqrt{2}$ means $+\sqrt{2}$. The numbers $-\sqrt{2}$ and $\sqrt{2}$ are different.*

Proof. Let $A = \{x \in \mathbf{R} : x^2 < 2\}$. A slight variation of Proposition 8.32(ii) is:

$$\text{if} \quad 0 < x \le y \quad \text{and} \quad 0 < z \le w \quad \text{then} \quad xz \le yw\,.$$

When we apply this with $x = y = w$ and $z = 1$, we deduce

$$\text{if} \quad x \ge 1 \quad \text{then} \quad x \le x^2\,.$$

Thus every element in A is bounded above by 2. Further, $1 \in A$, so A is nonempty. By Axiom 8.52, A has a least upper bound u, and so $\sqrt{2} = u$ is well defined. Note also that since $1 \in A$, $u \ge 1$.

We claim that $u^2 = 2$. To prove this, we will show that both (a) $u^2 > 2$ and (b) $u^2 < 2$ lead to contradictions.

(a) Suppose $u^2 > 2$. Let $\delta = \min\{1, u^2 - 2\}$ and $h = \frac{\delta}{4u}$; then

Note that u, h, and δ are all positive.

$$u^2 - (u-h)^2 = u^2 - u^2 - h^2 + 2uh = h(2u-h) < h \cdot 2u < \delta\,.$$

Thus the distance between $(u-h)^2$ and u^2 is less than δ, which in turn is less than or equal to the distance between 2 and u^2. But then

$$2 < (u-h)^2 < u^2\,.$$

Now we will prove that $u - h$ is an upper bound for A. Since $\delta \le 1$ and $u \ge 1$, $u - h > 0$. Thus $u - h$ is an upper bound for the set of negative members of A. If $x \in A$ is nonnegative, we have

$$x^2 < 2 < (u-h)^2,$$

and so Proposition 10.5 implies $x < u - h$ whenever $x \in A$. This proves that $u - h$ is an upper bound for A.

Thus we have a contradiction: $h > 0$ and $u - h$ is an upper bound for A, yet $u = \sup A$.

(b) Suppose $u^2 < 2$. Let $\delta = 2 - u^2$ and $h = \min\left\{\frac{\delta}{4u}, u\right\}$; then

$$(u+h)^2 - u^2 = u^2 + h^2 + 2uh - u^2 = h(2u+h) \le h \cdot 3u < \delta$$

and thus

$$u^2 < (u+h)^2 < 2\,.$$

By a similar argument as in (a), this implies $u + h \in A$, so u is not an upper bound for A, a contradiction. □

There is nothing special about the number 2 in the above definition and theorem. Given any $r \in \mathbf{R}_{>0}$, we define the **square root** of r by

$$\sqrt{r} := \sup\left\{x \in \mathbf{R} : x^2 < r\right\}.$$

Theorem 10.26. *Given any $r \in \mathbf{R}_{>0}$, the real number $\sqrt{r}$ is well defined, positive, and satisfies $\sqrt{r}^2 = r$.*

We also define $\sqrt{0} := 0$.

Proposition 10.27. *Given any $r \in \mathbf{R}_{>0}$, the number $\sqrt{r}$ is unique in the sense that, if x is a positive real number such that $x^2 = r$, then $x = \sqrt{r}$.*

Proposition 10.28. *If $r < 0$ there exists no $x \in \mathbf{R}$ such that $x^2 = r$.*

Proposition 10.28 expresses the fact that negative real numbers do not have square roots in $\mathbf{R}$. In Chapter C we will be studying a larger set of numbers than $\mathbf{R}$, called the *complex numbers*. Negative real numbers do have "square roots" in the complex numbers.

Review Question. Do you understand the (ε, N) definition of the limit of a sequence of real numbers?

Weekly reminder: Reading mathematics is not like reading novels or history. You need to think slowly about every sentence. Usually, you will need to reread the same material later, often more than one rereading.

This is a short book. Its core material occupies about 140 pages. Yet it takes a semester for most students to master this material. In summary: read line by line, not page by page.

Chapter 11
Rational and Irrational Numbers

5 out of 4 people have trouble with fractions.
Billboard in Danby, NY

Before You Get Started. What is a fraction? Is it a real number? Are all real numbers fractions? Have you seen fractions in this book up to now? What real number should the fraction $\frac{1}{2}$ be?

M. Beck and R. Geoghegan, *The Art of Proof: Basic Training for Deeper Mathematics*, Undergraduate Texts in Mathematics, DOI 10.1007/978-1-4419-7023-7_11,

11.1 Rational Numbers

The word rational *comes from* ratio. **Q** *stands for* quotient. *Much earlier in this book, you might have wondered about the origin of the symbol* **Z**. *It comes from the word* Zahlen, *which is* German *for* numbers.

A real number $z \in \mathbf{R}$ is **rational** if $z = \frac{m}{n}$, where $m, n \in \mathbf{Z}$ and $n \neq 0$. Nonrational real numbers are **irrational**. The set of all rational numbers is denoted by $\mathbf{Q}$. It is a subset of $\mathbf{R}$.

Irrational numbers will be discussed in the next section; in particular, we will see that not all real numbers are rational. For now, we develop fractions just as you have used them since elementary school.

Proposition 11.1. *$\mathbf{Z}$ is a subset of $\mathbf{Q}$.*

Proposition 11.2. *Let $x, y, z, w \in \mathbf{R}$ with $y \neq 0$ and $w \neq 0$. If $\frac{x}{y} = \frac{z}{w}$ then $xw = zy$.*

Proposition 11.3. *If $x, y, z \in \mathbf{R}$ with $y \neq 0$ and $z \neq 0$ then $\frac{xz}{yz} = \frac{x}{y}$.*

Hint: Theorem 6.32 is needed here.

Proposition 11.4. *Given a rational number $r \in \mathbf{Q}$, we can always write it as $r = \frac{m}{n}$, where $n > 0$ and m and n do not have any common factors.*

This representation of r is **in lowest terms**.

Proposition 11.5. *Let $m, n, s, t \in \mathbf{Z}$ be such that m and n do not have any common factors. If $\frac{m}{n} = \frac{s}{t}$ then m divides s and n divides t.*

By a prime power *we mean a prime number raised to a positive integer power.*

Proof. Assume $\frac{m}{n} = \frac{s}{t}$, where m and n do not have any common factors. Then $mt = sn$, and thus any prime power p^k appearing in the prime factorization of m has to appear in the prime factorization of sn. But m and n have no common (prime) factors, so p^k has to be part of the prime factorization of s. We have just proved that any prime power dividing m also has to divide s, which by uniqueness of prime factorizations (Theorem 6.32) implies that m divides s. The proof that n divides t is similar. □

You learned all this in elementary school under the title fractions. *As promised, we are systematically organizing some of the mathematics you previously knew.*

Proposition 11.6. *For all $m, n, s, t \in \mathbf{Z}$, where $n, t \neq 0$,*

$$\frac{m}{n} + \frac{s}{t} = \frac{mt + ns}{nt}.$$

Proposition 11.7. *The rational number $\frac{m}{n} \in \mathbf{Q}$ is positive (i.e., $\frac{m}{n} \in \mathbf{R}_{>0}$) if and only if either $m > 0$ and $n > 0$, or $m < 0$ and $n < 0$.*

In Theorem 8.43 we showed that between any two real numbers there is a third. We will now prove that between any two real numbers we can find a *rational* number.

Theorem 11.8. *Let $x, y \in \mathbf{R}$ with $x < y$. Then there exists a rational number r such that $x < r < y$.*

Proof. The case $x = 0$ is covered by Proposition 10.4. Next, we consider the case $x > 0$.

Proposition 11.17 will complement Theorem 11.8 by showing that there is also an irrational number between x and y.

Let $\varepsilon = y - x$; note that $\varepsilon > 0$. We need to prove that there exists $r \in \mathbf{Q}$ such that $x < r < x + \varepsilon$.

By Proposition 10.4, there exists $m \in \mathbf{N}$ such that $\frac{1}{m} < \varepsilon$. Furthermore, because $\mathbf{Z}$ is unbounded, there exists $n \in \mathbf{Z}$ such that $n > mx$. Let this n be minimal, that is, $n - 1 \le mx < n$. With Proposition 8.32, we can rewrite these inequalities as

$$\frac{n}{m} - \frac{1}{m} \le x < \frac{n}{m} \, .$$

The left-hand inequality implies $\frac{n}{m} \le x + \frac{1}{m} < x + \varepsilon$, so that together with the right-hand inequality, we deduce

$$x < \frac{n}{m} < x + \varepsilon = y \, .$$

Finally, if $x < 0$ we may assume $y < 0$ also (if not we can choose $r = 0$). If the rational number r satisfies $-y < r < -x$ (and there is such a number r since $-x > 0$) then $x < -r < y$. □

Corollary 11.9. *There is no smallest positive rational number.*

11.2 Irrational Numbers

Proposition 11.10. *The real number $\sqrt{2}$ is irrational.*

Proof. We will give a proof by contradiction. Assume that $\sqrt{2} = \frac{m}{n}$ for some $m, n \in \mathbf{Z}$. Because of Proposition 11.4, we can assume that m and n have no common factors. Now $2 = \frac{m^2}{n^2}$ implies that we can write

*This proof, due to Geoffrey C. Berresford, appeared in American Mathematical Monthly **115** (2008), p. 524.*

$$\frac{m}{n} = \frac{2n}{m} \, ,$$

and since $\frac{m}{n}$ is written in lowest terms, Proposition 11.5 implies that n divides m. But then $\frac{m}{n} = \sqrt{2}$ is an integer. We saw in the proof of Theorem 10.25 that $1 < \sqrt{2} < 2$. By Corollary 2.22, $\sqrt{2}$ cannot be an integer, so we have a contradiction. □

The next project implies that the Completeness Axiom 8.52 does not hold in $\mathbf{Q}$:

Project 11.11. Find a nonempty subset of $\mathbf{Q}$ that is bounded above but has no least upper bound in $\mathbf{Q}$. Justify your claim.

*Your set will have a least upper bound in **R**.*

An integer n is a **perfect square** if $n = m^2$ for some $m \in \mathbf{Z}$. You are invited to modify the proof of Proposition 11.10 to prove the following more general theorem.

Theorem 11.12. *If $r \in \mathbf{N}$ is not a perfect square, then $\sqrt{r}$ is irrational.*

Proposition 11.13. *Let m and n be nonzero integers. Then $\frac{m}{n}\sqrt{2}$ is irrational.*

Project 11.14. Prove that $\sqrt{2}+\sqrt{3}$ is irrational. Generalize.

Project 11.15. Here is the outline of an alternative proof of Proposition 11.10: Again, we suppose (hoping to obtain a contradiction) that $\sqrt{2} = \frac{m}{n}$ for some $m, n \in \mathbf{Z}$, and since we may write this fraction in lowest terms, m and n are not both even. Now $m^2 = 2n^2$, so m^2 is even, and Proposition 6.17 implies that m is even. So we can write $m = 2j$ for some integer j; then a quick calculation gives that $n^2 = 2j^2$, which means that n^2 is even. We can use Proposition 6.17 once more to deduce that n is even. But then both m and n are even, contrary to the first sentence of this proof. □
Compare the two proofs of Proposition 11.10. How do they differ? Are they really different? What are advantages/disadvantages of each?

We can also define higher roots: Namely, for an integer $n \geq 2$, the nth **root of** $r \in \mathbf{R}_{>0}$ is the positive real number $\sqrt[n]{r}$ that satisfies $(\sqrt[n]{r})^n = r$. Adapting the proof of Theorem 10.25, one can show that such a number exists in $\mathbf{R}$. The following proposition is analogous to Proposition 11.10.

Proposition 11.16. *The real number $\sqrt[n]{2}$ is irrational.*

Proposition 11.17. *Let $x, y \in R$ with $x < y$. There exists an irrational number z such that $x < z < y$.*

Hint: by Theorem 11.8, we can find $a, b \in \mathbf{Q}$ such that $x < a < b < y$. Now show that $a + \frac{b-a}{\sqrt{2}}$ is irrational and between x and y.

Corollary 11.18. *There is no smallest positive irrational number.*

If you can remember only a few things from this book, let the following be one of them:

> Between any two real numbers lies a rational number
> and also an irrational number.

Project 11.19. Try to plot the graph of the function $f : [0,1] \to \mathbf{R}$ given by

$$f(x) = \begin{cases} 0 & \text{if } x \in \mathbf{Q}, \\ 1 & \text{if } x \notin \mathbf{Q}. \end{cases}$$

The fact that the Completeness Axiom 8.52 does not hold in **Q** (see Project 11.11) is a major difference between **Q** and **R**. On the other hand, the following two propositions imply that **Q** and **R** have much in common, namely, **Q** satisfies Axioms 8.1–8.5 and 8.26 of Chapter 8.

Proposition 11.20. *Let $m,n,s,t,u,v \in \mathbf{Z}$, where $n,t,v \neq 0$.*

(i) *For all* $\frac{m}{n}, \frac{s}{t}, \frac{u}{v} \in \mathbf{Q}$:

  (a) $\frac{m}{n} + \frac{s}{t} = \frac{s}{t} + \frac{m}{n}$.

  (b) $\left(\frac{m}{n} + \frac{s}{t}\right) + \frac{u}{v} = \frac{m}{n} + \left(\frac{s}{t} + \frac{u}{v}\right)$.

  (c) $\frac{m}{n}\left(\frac{s}{t} + \frac{u}{v}\right) = \frac{m}{n} \cdot \frac{s}{t} + \frac{m}{n} \cdot \frac{u}{v}$.

  (d) $\frac{m}{n} \cdot \frac{s}{t} = \frac{s}{t} \cdot \frac{m}{n}$.

  (e) $\left(\frac{m}{n}\frac{s}{t}\right)\frac{u}{v} = \frac{m}{n}\left(\frac{s}{t}\frac{u}{v}\right)$.

(ii) *For all* $\frac{m}{n} \in \mathbf{Q}$, $\frac{m}{n} + 0 = \frac{m}{n}$.

(iii) *For all* $\frac{m}{n} \in \mathbf{Q}$, $\frac{m}{n} \cdot 1 = \frac{m}{n}$.

(iv) *For all* $\frac{m}{n} \in \mathbf{Q}$, $\frac{m}{n} + \frac{(-m)}{n} = 0$.

(v) *For all* $\frac{m}{n} \in \mathbf{Q} - \{0\}$, $\frac{m}{n} \cdot \frac{n}{m} = 1$.

Proposition 11.21.

(i) *The sum of two positive rational numbers is a positive rational number.*

(ii) *The product of two positive rational numbers is a positive rational number.*

(iii) *For every* $\frac{m}{n} \in \mathbf{Q}$ *such that* $\frac{m}{n} \neq 0$, *either* $\frac{m}{n}$ *is positive or* $\frac{-m}{n}$ *is positive, and not both.*

11.3 Quadratic Equations

A (real) **quadratic equation** is an equation of the form $ax^2 + bx + c = 0$ where $a,b,c \in \mathbf{R}$ with $a \neq 0$.

Here we have to demand that $a \neq 0$ for the equation $ax^2 + bx + c = 0$ to be called quadratic. An equation of the form $bx + c = 0$ with $b \neq 0$ is called a linear equation.

Proposition 11.22. *If $a,b,c \in \mathbf{R}$, where a and b are not both zero, then the equation $ax^2 + bx + c = 0$ has a solution $x \in \mathbf{R}$ if and only if $b^2 - 4ac \geq 0$.*

In fact, as you well know, $\frac{-b+\sqrt{b^2-4ac}}{2a}$ is a solution (assuming that $a \neq 0$); and $\frac{-b-\sqrt{b^2-4ac}}{2a}$ is also a solution. The number $b^2 - 4ac$ is called the **discriminant** of the equation $ax^2 + bx + c = 0$.

Corollary 11.23. *The equation $x^2 + 1 = 0$ does not have a solution in* $\mathbf{R}$.

Project 11.24. How many solutions can a quadratic equation have? Justify your claims.

Proposition 11.25. *Let $b, c, p, q \in \mathbf{R}$. If $x^2 - bx - c = 0$ has the two solutions s and t, and if we define a sequence $(a_k)_{k=1}^{\infty}$ by*

$$a_k := ps^k + qt^k,$$

then this sequence satisfies the recurrence relation

$$a_n = ba_{n-1} + ca_{n-2} \qquad \text{for all } n \geq 3.$$

Setting $b = c = 1$, we get the recurrence relation $a_n = a_{n-1} + a_{n-2}$, which (when we start with $a_1 = a_2 = 1$) is the defining recurrence relation for the Fibonacci numbers, which we studied in Section 4.6. Thus the quadratic equation $x^2 - x - 1 = 0$ has some connection with the Fibonacci numbers.

Project 11.26. Prove that the kth Fibonacci number is given by

$$f_k = \frac{1}{\sqrt{5}} \left(\left(\frac{1+\sqrt{5}}{2} \right)^k - \left(\frac{1-\sqrt{5}}{2} \right)^k \right).$$

Compare your proof with the one we have given for Proposition 4.29.

Review Questions. Do you understand that between any two real numbers there lies both a rational number and an irrational number? Do you see that there is neither a smallest positive real number nor a smallest positive rational number?

Weekly reminder: Reading mathematics is not like reading novels or history. You need to think slowly about every sentence. Usually, you will need to reread the same material later, often more than one rereading.

This is a short book. Its core material occupies about 140 pages. Yet it takes a semester for most students to master this material. In summary: read line by line, not page by page.

Chapter 12
Decimal Expansions

FOXTROT © Bill Amend. Reprinted with permission of UNIVERSAL UCLICK.

Our goal in this chapter is to prove that every real number can be represented by a decimal and that every decimal represents a real number.

Before You Get Started. In Chapter 7 we discussed base-ten representation of integers. In fact, all real numbers can be given base-ten representations: this is what you called "decimals" in grade school. Do you remember repeating and nonrepeating decimals? For example, the decimal expansion $0.1111\ldots$ is supposed to represent the number $\frac{1}{9}$. Why is this a valid expression for $\frac{1}{9}$? What is a decimal expansion of $\frac{1}{8}$ or $\sqrt{2}$?

M. Beck and R. Geoghegan, *The Art of Proof: Basic Training for Deeper Mathematics*, Undergraduate Texts in Mathematics, DOI 10.1007/978-1-4419-7023-7_12,

12.1 Infinite Series

We saw in Chapter 4 that the notation $\sum_{j=1}^{k} a_j$ is used for the "sum" obtained when the numbers $a_1, a_2, \ldots, a_k$ are added together. One frequently uses the less precise notation $a_1 + a_2 + \cdots + a_k$ for that same "sum." Here we want to discuss the meaning of $\sum_{j=1}^{\infty} a_j$, which can be written less formally as $a_1 + a_2 + \cdots$. The notation suggests that we are adding infinitely many numbers and that we have a notation for their "sum."

Consider the difficulties hidden in writing an infinite sum:

(i) $1+1+1+\cdots$; i.e., every number a_j is equal to 1.

(ii) $1-1+1-1+1-1+\cdots$; i.e., $a_j = (-1)^{j+1}$.

(iii) $1+\frac{1}{2}+\frac{1}{4}+\frac{1}{8}+\cdots$; i.e., $a_j = \frac{1}{2^{j-1}}$.

(iv) $1+\frac{1}{2}+\frac{1}{3}+\frac{1}{4}+\cdots$; i.e., $a_j = \frac{1}{j}$.

(v) $1+\frac{1}{4}+\frac{1}{9}+\frac{1}{16}+\cdots$; i.e., $a_j = \frac{1}{j^2}$.

In item (i), adding up infinitely many 1's does not give a finite answer. In item (ii), alternately adding 1 and -1 does not look promising. Item (iii) is what we call a geometric series, and you probably know that its "sum" is considered to be 2. Items (iv) and (v) should be familiar from calculus: (iv) "diverges," while the "sum" in (v) is considered to be $\frac{\pi^2}{6}$.

This informal discussion raises the following question: what exactly does all this mean? In particular, what is meant by "is considered to be" and "diverges" and "does not look promising"?

We begin again.

An expression of the form $\sum_{j=1}^{\infty} a_j$, where each a_j is a real number, is called a **series** or an **infinite series**. Hidden in this definition are several items.

- There is a sequence of real numbers $(a_j)_{j=1}^{\infty}$, called the **sequence of terms** of the series.
- There is another sequence of numbers $(s_k)_{k=1}^{\infty}$ formed from the sequence of terms by the formula $s_k = \sum_{j=1}^{k} a_j$; the number s_k is called the kth **partial sum** of the series, and the sequence $(s_k)_{k=1}^{\infty}$ is the **sequence of partial sums**.
- If the sequence of partial sums converges to the number L, then L is called the **sum** of the series, and one writes

$$\sum_{j=1}^{\infty} a_j = L.$$

- If the series has a sum, it is said to **converge**; if the series has no sum (i.e., if the sequence of partial sums diverges) the series is said to **diverge**.

Example 12.1. Given two real numbers a and x the series whose jth term is ax^j is called the **geometric series** with 0th term a and **common ratio** x. Consider the case $a = 1$ and $x = \frac{1}{4}$. We saw in Proposition 4.13 that the kth partial sum is

$$\sum_{j=0}^{k} \frac{1}{4^j} = \frac{1 - \left(\frac{1}{4}\right)^{k+1}}{1 - \frac{1}{4}} = \frac{4}{3}\left(1 - \left(\tfrac{1}{4}\right)^{k+1}\right).$$

Since you proved in Proposition 10.18 that $\lim_{k\to\infty} \left(\frac{1}{4}\right)^{k+1} = 0$, we see that in the limit,

$$\sum_{j=0}^{\infty} \frac{1}{4^j} = \frac{4}{3}.$$

It is essentially this limit that is illustrated on the cover of this book: $\frac{1}{4} + \frac{1}{4^2} + \frac{1}{4^3} + \cdots = \frac{1}{3}$.

This example is the special case $a = 1$, $x = \frac{1}{4}$ of the following proposition.

Proposition 12.2. *When $|x| < 1$, the geometric series $\sum_{j=0}^{\infty} ax^j$ converges to $\frac{a}{1-x}$; i.e.,*

$$\sum_{j=0}^{\infty} ax^j = \frac{a}{1-x}.$$

Proof. By (the $\mathbf{R}$-analogues of) Propositions 4.15(i) and 4.13, we have for $x \neq 1$ and $k \in \mathbf{Z}_{\geq 0}$,

$$\sum_{j=0}^{k} ax^j = a\sum_{j=0}^{k} x^j = a\,\frac{1 - x^{k+1}}{1 - x},$$

so by Proposition 10.23 we have only to show that

$$\lim_{k\to\infty} x^{k+1} = \lim_{k\to\infty} x^k = 0$$

when $|x| < 1$. But this is the content of Proposition 10.18. □

Our next goal is to prove the following *Comparison Test*:

Proposition 12.3. *Let $0 \leq a_k \leq b_k$ for all $k \geq 0$. If $\sum_{j=0}^{\infty} b_j$ converges then $\sum_{j=0}^{\infty} a_j$ converges.*

Proof. Let $0 \leq a_k \leq b_k$ for all $k \geq 0$, and let

$$L = \sum_{j=0}^{\infty} b_j.$$

Since $b_k \geq 0$, the sequence $(B_k)_{k=0}^{\infty}$ of partial sums $B_k := \sum_{j=0}^{k} b_j$ is increasing, and by Proposition 10.21, $B_k \leq L$ for all $k \geq 0$. Let $A_k := \sum_{j=0}^{k} a_j$. Then for all $k \geq 0$,

$$0 \le A_k = \sum_{j=0}^{k} a_j \le \sum_{j=0}^{k} b_j = B_k \le L,$$

so the sequence $(A_k)_{k=0}^{\infty}$ of partial sums is bounded. Since $a_k \ge 0$, $(A_k)_{k=0}^{\infty}$ is also increasing, and so by Theorem 10.19, $\sum_{j=0}^{\infty} a_j$ converges. □

12.2 Decimals

A **nonnegative decimal** is a sequence $(m, d_1, d_2, d_3, \dots)$ where $m \ge 0$ is an integer and each d_n is a digit, that is, an integer between 0 and 9. By tradition (as you well know) the notation used for a nonnegative decimal is $m.d_1 d_2 d_3 \dots$. This nonnegative decimal **represents** the real number

As you know, in practice we will write the integer m in base-ten representation as discussed in Chapter 7.

$$x = m + \sum_{j=1}^{\infty} d_j 10^{-j}.$$

This number x is nonnegative. We call $(m, d_1, d_2, d_3, \dots)$ a **decimal expansion** of x. A decimal expansion of a negative real number x is defined by placing a minus sign in front of a decimal expansion of the positive number $-x$.

For all of this to make sense, we need the following result.

Proposition 12.4. *Let* $(d_k)_{k=1}^{\infty}$ *be a sequence of digits. Then* $\sum_{j=1}^{\infty} d_j 10^{-j}$ *converges.*

Proposition 12.5. *Let* $(d_k)_{k=1}^{\infty}$ *be a sequence of digits and* $n \in \mathbf{N}$. *Then*

$$\sum_{j=n}^{\infty} d_j 10^{-j} \le \frac{1}{10^{n-1}}.$$

Proposition 12.4 implies that every decimal expansion represents a real number. Now we will prove the converse:

Theorem 12.6. *Every real number has a decimal expansion.*

Proof. We will prove this theorem for nonnegative real numbers. The general case follows easily: if $x < 0$, just get a decimal expansion of $-x$ and precede it with a minus sign.

So let $x \ge 0$. We will recursively define a decimal expansion $m.d_1 d_2 d_3 \dots$ of x. Let m be the smallest integer for which $x < m + 1$. Then $m \le x$ (otherwise m was not chosen minimally). Next, let d_1 be the smallest element of

We are using the Well-Ordering Principle (Theorem 2.32) here.

$$\left\{ n \in \mathbf{Z}_{\ge 0} : \ x < m + \frac{n+1}{10} \right\}.$$

Then $0 \le d_1 \le 9$ (otherwise m was not chosen minimally) and $m + \frac{d_1}{10} \le x$ (otherwise d_1 was not chosen minimally). In summary,

$$m + \frac{d_1}{10} \le x < m + \frac{d_1 + 1}{10} .$$

Now we define the remaining digits recursively. Assuming that $d_1, d_2, \ldots, d_k$ have been defined so that

$$m + \sum_{j=1}^{k} d_j 10^{-j} \le x < m + \sum_{j=1}^{k-1} d_j 10^{-j} + \frac{d_k + 1}{10^k} ,$$

let d_{k+1} to be the smallest element of

$$\left\{ n \in \mathbf{Z}_{\ge 0} : \ x < m + \sum_{j=1}^{k} d_j 10^{-j} + \frac{n+1}{10^{k+1}} \right\}.$$

Then

$$m + \sum_{j=1}^{k+1} d_j 10^{-j} \le x < m + \sum_{j=1}^{k} d_j 10^{-j} + \frac{d_{k+1} + 1}{10^{k+1}} .$$

This recursive definition ensures that for $k \in \mathbf{N}$,

$$0 \le x - \left(m + \sum_{j=1}^{k} d_j 10^{-j} \right) < \frac{1}{10^k} ,$$

which will allow us to prove that

$$x = m + \sum_{j=1}^{\infty} d_j 10^{-j} .$$

Namely, for a given $\varepsilon > 0$, by Proposition 10.4, there exists $N > \frac{1}{\varepsilon}$; then for $n \ge N$,

$$\left| x - \left(m + \sum_{j=1}^{n} d_j 10^{-j} \right) \right| = x - \left(m + \sum_{j=1}^{n} d_j 10^{-j} \right) < \frac{1}{10^n} < \frac{1}{n} \le \frac{1}{N} < \varepsilon .$$

Here we are using Proposition 7.1.

This means that the partial sums $m + \sum_{j=1}^{k} d_j 10^{-j}$ converge to x as $k \to \infty$, which is what we set out to prove. □

Next, we consider the uniqueness question: can a real number have more than one decimal expansion, and if so how many? We start with the special case of the real number 1.

Proposition 12.7. *Let $m.d_1 d_2 d_3 \ldots$ represent $1 \in \mathbf{R}$. Then either $m = 1$ and every d_k equals 0, or $m = 0$ and every d_k equals 9. In other words, 1 can be represented by $1.00000\ldots$ and $0.999999\ldots$, and by no other decimal.*

Proof. The expansion $1.00000\cdots = 1+\sum_{j=1}^{\infty} 0\cdot 10^{-j}$ certainly represents 1, as does

$$0.999999\cdots = \sum_{j=1}^{\infty} 9\cdot 10^{-j} = 9\sum_{j=1}^{\infty}\left(\frac{1}{10}\right)^j = 9\left(\frac{1}{1-\frac{1}{10}}-1\right),$$

by Proposition 12.2. We must show that there are no other decimal expansions of 1.

So let $m+\sum_{j=1}^{\infty}\frac{d_j}{10^j}$ be a decimal expansion of 1. If $m\geq 2$ or $m\leq -1$ then this expansion differs from 1 by at least 1, so we just have to consider the cases $m=0$ and $m=1$.

Case 1: $m=0$. Let N be the smallest subscript $n\geq 1$ for which $d_n<9$. (If all d_n equal 9 we get the expansion $0.999999\ldots$.) Then

$$\sum_{j=1}^{N}\frac{d_j}{10^j} = \sum_{j=1}^{N-1}\frac{9}{10^j}+\frac{d_N}{10^N} = 9\left(\frac{1-10^{-N}}{1-\frac{1}{10}}-1\right)+\frac{d_N}{10^N},$$

by Proposition 4.13. The expression on the right-hand side simplifies to

$$\sum_{j=1}^{N}\frac{d_j}{10^j} = 1-\frac{1}{10^{N-1}}+\frac{d_N}{10^N} = 1-\frac{10-d_N}{10^N}.$$

But then

$$1-\sum_{j=1}^{\infty}\frac{d_j}{10^j} = \frac{10-d_N}{10^N} - \sum_{j=N+1}^{\infty}\frac{d_j}{10^j}.$$

Since $d_N<9$, the first term on the right-hand side is at least $\frac{2}{10^N}$. The second term is bounded above by $\frac{1}{10^N}$, by Proposition 12.5. Hence

$$1-\sum_{j=1}^{\infty}\frac{d_j}{10^j}\geq\frac{1}{10^N},$$

and Proposition 10.11 implies that $\sum_{j=1}^{\infty}\frac{d_j}{10^j}\neq 1$.

Case 2: $m=1$. Let N be the smallest subscript $n\geq 1$ for which $d_n>0$. (If all d_n are 0 we get the expansion $1.000000\ldots$.) Then

$$1+\sum_{j=1}^{N}\frac{d_j}{10^j} = 1+\frac{d_N}{10^N}.$$

But then

$$\left(1+\sum_{j=1}^{\infty}\frac{d_j}{10^j}\right)-1 = \frac{d_N}{10^N}+\sum_{j=N+1}^{\infty}\frac{d_j}{10^j}\geq\frac{1}{10^N},$$

since $d_N > 0$. Again Proposition 10.11 implies that $1 + \sum_{j=1}^{\infty} \frac{d_j}{10^j} \neq 1$. □

The above proof contains all the necessary ingredients for the following more general theorem.

Theorem 12.8. *Let $m.d_1d_2d_3\ldots$ and $n.e_1e_2e_3\ldots$ be different decimal expansions of the same nonnegative real number.*

(i) *If $m < n$, then $n = m+1$, every e_k is 0 and every d_k is 9.*

(ii) *If $m = n$, let N denote the smallest subscript such that $d_N \neq e_N$. If $d_N < e_N$ then $e_N = d_N + 1$, $e_j = 0$ for all $j > N$, and $d_j = 9$ for all $j > N$.*

Thus if a real number has two different decimal expansions then one of those expansions has only finitely many nonzero digits. This implies the following corollary.

Corollary 12.9. *If $r \in \mathbf{R}$ has two different decimal expansions then r is a rational number.*

A nonnegative decimal $(m, d_1, d_2, d_3, \ldots)$ is **repeating** if there exist $N \in \mathbf{N}$ and $p \in \mathbf{N}$ such that for all $0 \leq n < p$ and for all $k \in \mathbf{N}$,

$$d_{N+n+kp} = d_{N+n}\,.$$

The simplest example of a repeating decimal is one with only finitely many digits (i.e., the repeating digits are zeros).

Project 12.10. Show that $5.666\ldots$ and $0.34712712712712\ldots$ are rational. Once you have understood these two examples, prove the following theorem.

Hint: There are geometric series hidden here.

Theorem 12.11. *Every repeating decimal represents a rational number.*

The converse is also true:

Project 12.12. Express $\frac{71}{13}$ and $\frac{34}{31}$ as decimals. Once you have understood these two examples, prove the following theorem.

Hint: use the division algorithm (Theorem 6.13).

Theorem 12.13. *Every rational number is represented by a repeating decimal.*

Review Questions. Do you understand that when you studied decimals you were really studying real numbers? Which real numbers have two different decimal expressions? Why does a real number never have three different decimal expressions?

Weekly reminder: Reading mathematics is not like reading novels or history. You need to think slowly about every sentence. Usually, you will need to reread the same material later, often more than one rereading.

This is a short book. Its core material occupies about 140 pages. Yet it takes a semester for most students to master this material. In summary: read line by line, not page by page.

Chapter 13
Cardinality

Sometimes these cogitations still amaze
The troubled midnight, and the noon's repose.
T. S. Eliot (*La Figlia Che Piange*)

FRAZZ: © Jef Mallett / Dist. by United Feature Syndicate, Inc. Reprinted with permission.

Before You Get Started. The goal of this chapter is to compare the sizes of infinite sets. It is perfectly sensible to say that the sets $\{1,2,4\}$ and $\{2,3,5\}$ have the same size (still you might think about how to define rigorously what it means for two finite sets to have equal size), but how do we compare the sizes of $\mathbf{N}$ and $\mathbf{Z}$, of $\mathbf{N}$ and $\mathbf{R}_{>0}$, of $\mathbf{Q}$ and $\mathbf{R}$? We want to say more than just "they are all infinite"—how do they compare? Are they all of different sizes? More generally, how should we measure the size of infinite sets?

M. Beck and R. Geoghegan, *The Art of Proof: Basic Training for Deeper Mathematics*, Undergraduate Texts in Mathematics, DOI 10.1007/978-1-4419-7023-7_13,

13.1 Injections, Surjections, and Bijections Revisited

Here is what we mean when we say that two sets "have the same size": The sets A and B **have the same cardinality** (or **have the same cardinal number**) if there exists a bijection $A \longrightarrow B$.

A special case is that of finite sets, for example, $\{1,2,\ldots,n\}$ for some $n \in \mathbf{N}$. Since we will use this set frequently in this chapter, we denote $\{1,2,\ldots,n\}$ by $[n]$.

A set S is **finite** if either $S = \varnothing$ or for some $n \in \mathbf{N}$ there exists a bijection from $[n]$ to S. An **infinite** set is one that is not finite. A set S is **countably infinite** if there exists a bijection from $\mathbf{N}$ to S. A set S is **countable** if either S is finite or S is countably infinite.

It may seem obvious that there is no bijection $[m] \longrightarrow [n]$ when $m \neq n$, but it needs proof (Theorem 13.4) and is not trivial. The steps needed for the proof are given here as Propositions 13.1–13.3.

Proposition 13.1. *There exists no bijection* $[1] \longrightarrow [n]$ *when* $n > 1$.

In other words, f and $\tilde{f}$ are defined by the same rule, but they have different domains and codomains.

Proposition 13.2. *If* $f : A \longrightarrow B$ *is a bijection and* $a \in A$*, define the new function*

$$\tilde{f} : A - \{a\} \longrightarrow B - \{f(a)\} \qquad \text{by} \qquad \tilde{f}(x) := f(x).$$

Then $\tilde{f}$ *is well defined and bijective.*

Proposition 13.3. *If* $1 \leq k \leq n$ *then the function*

$$g_k : [n-1] \longrightarrow [n] - \{k\} \qquad \text{defined by} \qquad g_k(j) := \begin{cases} j & \text{if } j < k, \\ j+1 & \text{if } j \geq k \end{cases}$$

is a bijection.

Theorem 13.4. *Let* $m, n \in \mathbf{N}$. *If* $m \neq n$*, there exists no bijection* $[m] \longrightarrow [n]$.

Proof. We prove the statement $P(m)$:

$$\text{For all } n \neq m \text{ there exists no bijection } [m] \longrightarrow [n]$$

by induction on $m \in \mathbf{N}$. The base case is captured by Proposition 13.1.

For the induction step, let $m \geq 2$ and assume we know that $P(m-1)$ is true. Suppose (by way of contradiction) there exists a bijection $f : [m] \longrightarrow [n]$ for some $n \neq m$. Let $k = f(m)$ and define, as in Proposition 13.2,

$$\tilde{f} : [m-1] \longrightarrow [n] - \{k\} \qquad \text{by} \qquad \tilde{f}(x) := f(x).$$

Proposition 13.2 says that this new function is also bijective. The composition of $\tilde{f}$ with the inverse of the function g_k defined in Proposition 13.3 gives a bijection $[m-1] \to [n-1]$, by Proposition 9.7. But this contradicts our induction hypothesis. □

Thus for finite sets the **number of elements** is well defined: a set S **contains** n **elements** if and only if there exists a bijection from $[n]$ to S. Then every set having the same cardinal number as S also contains n elements. We say that $\varnothing$ contains 0 elements.

Proposition 13.5 (Pigeonhole Principle). *If $m > n$ then a function $[m] \to [n]$ cannot be injective.*

Proposition 13.5 implies that if $m > n$ and we label n objects with numbers from 1 to m then there exist two objects that have the same label. The Pigeonhole Principle appears in many different areas in mathematics and beyond. It asserts that if there are n pigeonholes and m pigeons, there are at least two pigeons who must share a hole; or if there are n people in an elevator and m buttons are pressed, someone is playing a practical joke.

Proposition 13.6. *Every subset of a finite set is finite.*

Proposition 13.7. *A nonempty subset of* **N** *is finite if and only if it is bounded above.*

Proposition 13.8. **N** *is infinite.*

Proposition 13.9. *The nonempty set A is countable if and only if there exists a surjection* $\mathbf{N} \to A$.

Proof. Assume A is nonempty and countable.

If A is finite, then there exist $n \in \mathbf{N}$ and a bijection $\phi : [n] \to A$. Let $\psi : \mathbf{N} \to [n]$ be defined by

$$\psi(m) := \begin{cases} m & \text{if } 1 \le m \le n, \\ 1 & \text{otherwise.} \end{cases}$$

This function ψ is surjective, and so by Proposition 9.7, $\phi \circ \psi : \mathbf{N} \to A$ is a surjection.

If A is infinite, then there exists a bijection from **N** to A, which is certainly surjective.

Conversely, assume there exists a surjection $\sigma : \mathbf{N} \to A$. If A is finite, it is countable by definition, and we are done.

If A is infinite, we define a bijection $\beta : \mathbf{N} \to A$ recursively as follows: $\beta(1) = \sigma(1)$, and for $n \ge 2$,

$$\beta(n) = \sigma(m_n),$$

The set on the right-hand side is not empty because A is infinite.

where

$$m_n := \min\{k \in \mathbf{N} : \sigma(k) \notin \{\sigma(1), \sigma(2), \ldots, \sigma(n-1)\}\}. \qquad \square$$

Proposition 13.10. *A subset of a countable set is countable.*

Proposition 13.11. *Every infinite set contains an infinite subset that is countable.*

Theorem 13.12. *A set is infinite if and only if it contains a proper subset that is also infinite.*

13.2 Some Countable Sets

The next propositions are counterintuitive at first sight.

Proposition 13.13. $\mathbf{Z}$ *is countable.*

Theorem 13.14. $\mathbf{Z} \times \mathbf{Z}$ *is countable.*

Here is the idea of the proof:

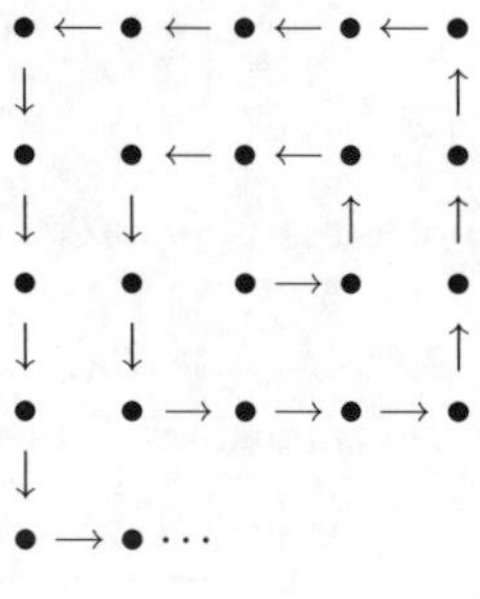

Project 13.15. Find an explicit formula for a bijection between $\mathbf{N}$ and $\mathbf{Z} \times \mathbf{Z}$.

Corollary 13.16. $\mathbf{N} \times \mathbf{N}$ *is countable.*

Corollary 13.17. $\mathbf{Z} \times (\mathbf{Z} - \{0\})$ *is countable.*

Corollary 13.18. $\mathbf{Q}$ *is countable.*

We know that $\mathbf{N} \subseteq \mathbf{Z} \subseteq \mathbf{Q}$; moreover, each is a proper subset of the next one, i.e., $\mathbf{N} \neq \mathbf{Z}$ and $\mathbf{Z} \neq \mathbf{Q}$. This might make you think that $\mathbf{N}$ is smaller than $\mathbf{Z}$ and that $\mathbf{Z}$ is smaller than **Q**. But we have just proved that these sets have the same cardinality—i.e., they have the same size. This can be confusing for beginners: if A and B are finite sets and A is a proper subset of B, then A and B have different cardinality by Theorem 13.4. We are seeing here that no such statement holds for infinite sets.

Proposition 13.19. *The countable union of countable sets is countable, i.e., if A_n is a countable set for each $n \in \mathbf{N}$ then $\bigcup_{n=1}^{\infty} A_n$ is countable.*

A real number is **algebraic** if it is the root of a polynomial with integer coefficients. A real number that is not algebraic is called **transcendental**.

Example 13.20. Every rational number is algebraic. The irrational numbers $\sqrt{3}$ and $\sqrt[4]{2}$ are algebraic. The numbers π and e, which you have studied in trigonometry and calculus, are transcendental (but this is not easy to prove).

Can you see why $r \in \mathbf{Q}$ and $\sqrt{3}$ are algebraic?

Proposition 13.21. *The set of algebraic numbers is countable.*

Hint: use Proposition 6.20.

13.3 Some Uncountable Sets

Theorem 13.22. **R** *is not countable.*

Proof. We will prove this by contradiction. Suppose that **R** is countable, and so by Proposition 13.9 there exists a surjective function $f : \mathbf{N} \to \mathbf{R}$. By Theorem 12.8, every real number has at most two decimal representations. So for each $n \in \mathbf{N}$, $f(n)$ can be written in the form $\pm m^{(n)}.d_1^{(n)} d_2^{(n)} d_3^{(n)} \ldots$; if there is more than one such decimal for $f(n)$, we use the one that has infinitely many nonzero digits.

Now let y be the real number represented by $0.a_1a_2a_3\ldots$, where

$$a_n := \begin{cases} 3 & \text{if } d_n^{(n)} \neq 3, \\ 4 & \text{if } d_n^{(n)} = 3. \end{cases}$$

Then for all $n \in \mathbf{N}$, $y \neq f(n)$, because the nth decimal places of y and $f(n)$ do not agree. Hence $y \in \mathbf{R}$ is not in the image of f, which contradicts the fact that f is surjective. □

This diagonalization argument *is due to Georg Cantor (1845–1928). Its appearance (Theorem 13.22) was the first proof that infinite sets may have different cardinal numbers.*

It follows that there is no one-to-one correspondence between the infinite sets **R** and **Q**, i.e., no function $f : \mathbf{Q} \to \mathbf{R}$ that is bijective. In particular, the "inclusion function" $\mathbf{Q} \to \mathbf{R}$ that takes each rational number to itself (regarded as a real number) is not surjective. This gives another proof that there exist irrational real numbers.

The discovery that **R** and **Q** have different cardinality, i.e., different size, was considered revolutionary in the mathematics of the late nineteenth century. It was not

Cantor originally conjectured the Continuum Hypothesis*: there is no uncountable set whose cardinality is smaller than that of* **R**. *Kurt Gödel (1906–1978) and Paul Cohen (1934–2007) proved that the usual axioms of set theory do not imply or refute the Continuum Hypothesis.*

that people had thought the opposite to be true; they just had never seriously considered the idea of infinite sets having different sizes. The foundations of the part of mathematics called *analysis* had to be completely rethought because of this.

Corollary 13.23. *The set* $\mathbf{R}-\mathbf{Q}$ *of irrational numbers is uncountable.*

Corollary 13.24. *The set of transcendental numbers is uncountable.*

The proof of Theorem 13.22 reveals even more. It shows that the set of decimals

$$\{0.d_1d_2d_3\cdots : \text{each } d_j = 3 \text{ or } 4\} \subseteq \mathbf{R}$$

is uncountable. Consequently, the interval $[0,1] = \{x \in \mathbf{R} : 0 \le x \le 1\}$ is uncountable. This construction can be modified to prove the following theorem.

A similar statement holds for open and half-open intervals.

Theorem 13.25. *Every interval* $[x,y]$ *is uncountable.*

This gives us a new proof of Proposition 11.17:

Corollary 13.26. *Let* $x,y \in \mathbf{R}$ *with* $x < y$. *Then there exists an irrational number* z *such that* $x < z < y$.

This proof establishes the existence of irrational numbers in all intervals without explicitly describing any, in contrast with our hint for Proposition 11.17.

Proof. In any interval, there are only countably many rational numbers, so there must be an irrational number. □

Corollary 13.27. *Between any two real numbers lies and algebraic number and also a transcendental number.*

Every interval is uncountable, and **R** has larger cardinality than **N**. A natural question is, where does the cardinality of an interval fit into this picture? Here is the answer:

Theorem 13.28. *Every open interval* (a,b) *has the same cardinality as* **R**.

Corollary 13.29. *All open intervals have the same cardinality.*

13.4 An Infinite Hierarchy of Infinities

Note that we use the card *symbol only when comparing the cardinalities of two sets. We discuss cardinal numbers further in Chapter F.*

We write $\operatorname{card} A \le \operatorname{card} B$ if there exists an injection $A \to B$. By Proposition 9.12, $\operatorname{card} A \le \operatorname{card} B$ is equivalent to saying that there exists a surjection $B \to A$. We write $\operatorname{card} A = \operatorname{card} B$ if A and B have the same cardinality, i.e., if there exists a bijection $A \to B$. We write $\operatorname{card} A < \operatorname{card} B$ when $\operatorname{card} A \le \operatorname{card} B$ and $\operatorname{card} A \ne \operatorname{card} B$.

If A is a set, let $P(A)$ denote the set containing all subsets of A, called the **power set** of A.

Example 13.30. If $A = \{a,b\}$, $P(A) = \{\varnothing, \{a\}, \{b\}, \{a,b\}\}$.

In this example, A has 2 members, and $P(A)$ has 4 members. Our goal in this section is to prove that the power set of A is always "bigger" than A:

Theorem 13.31. *For every set A,* $\operatorname{card} A < \operatorname{card} P(A)$.

Theorem 13.31 is profound. It implies an infinite hierarchy of infinities. For example, it says that $P(\mathbf{N})$ is not countable, $P(P(\mathbf{N}))$ has larger cardinality than $P(\mathbf{N})$, $P(P(P(\mathbf{N})))$ is yet larger than $P(P(\mathbf{N}))$, etc.

We start the proof of Theorem 13.31 with the finite case, for which we can be more precise:

Proposition 13.32. *For each* $n \in \mathbf{N}$, $\operatorname{card} P([n]) = \operatorname{card}[2^n]$.

Before proving Proposition 13.32, think through the case $n = 1$.

Proof of Theorem 13.31. The injection $\iota : A \to P(A)$, $\iota(x) = \{x\}$ shows that $\operatorname{card} A \leq \operatorname{card} P(A)$; we show that there is no *surjection* $A \to P(A)$.

Suppose, by way of contradiction, that $f : A \to P(A)$ is surjective. Let

$$B := \{a \in A : a \notin f(a)\}.$$

Compare this argument with Cantor diagonalization in our proof of Theorem 13.22.

This set B is an element of $P(A)$, so by our assumption, there exists $c \in A$ such that $f(c) = B$.

If $c \in B$ then $c \in f(c)$, and so $c \notin B$.

If $c \notin B$ then $c \notin f(c)$, and so $c \in B$.

Either way we arrive at a contradiction. □

Cardinality questions are often difficult to answer. For example, the Cantor–Schröder–Bernstein Theorem asserts that if $\operatorname{card} A \leq \operatorname{card} B$ and $\operatorname{card} B \leq \operatorname{card} A$ then $\operatorname{card} A = \operatorname{card} B$. Another important theorem says that if A and B are sets, then either $\operatorname{card} A \leq \operatorname{card} B$ or $\operatorname{card} B \leq \operatorname{card} A$. These are proved in Chapter F.

13.5 Nondescribable Numbers

This section is somewhat nonrigorous but is intended to make you aware of how elusive real numbers are.

We have proved that the set **R** of real numbers is uncountable. There is a sense in which many members of **R** cannot be described at all. We have already referred to algorithms and have admitted that defining the word "algorithm" is complicated.

Here, we will simply say that an algorithm is a finite set of mathematical procedures that when applied to an "input" produces an "output."

As an example, suppose you input the question: "find 313 times 498" into your calculator. The answer 155874 will appear on the screen: this is the output. Programmed into your calculator is a multiplication algorithm that works on your input and gives the output. If you use a large and powerful computer rather than a calculator, you can input huge numbers and get huge outputs, though it may take some time for the computer to execute all the instructions given by the algorithm. The important point to note is that the algorithm's length is not related to the size of the numbers you input, but rather the number of steps in applying the algorithm to your input will depend on the size of the input. The algorithm is a program containing loops, and the computer uses these loops as often as necessary for a particular input.

The question is this: given a real number x, is there an algorithm that will print out as many decimal places of x as you desire? In other words, will the algorithm print out arbitrarily many decimal places?

What might we mean by "given a real number"?

Here π is the ratio of the circumference of a circle to its diameter. This formula can be derived from the Taylor series of the function arctan.

Example 13.33. You learned in calculus that

$$\frac{\pi}{4} = 1 - \frac{1}{3} + \frac{1}{5} - \frac{1}{7} + \cdots .$$

Thus the sequence of partial sums $\left(\sum_{j=1}^{k}(-1)^{j+1}\frac{1}{2j-1}\right)_{k=1}^{\infty}$ converges to $\frac{\pi}{4}$, and so the sequence of partial sums of

$$\sum_{j=1}^{\infty}(-1)^{j+1}\frac{4}{2j-1}$$

converges to π. An algorithm that computes the decimal expansion of these partial sums can print out arbitrarily many decimal places of π. You can say how accurate you want the answer to be; this can easily be translated into what partial sum must be calculated, and the algorithms built in to the computer will do the rest.

Example 13.34. We defined $\sqrt{2} = \sup\{x \in \mathbf{R} : x^2 < 2\}$. Thus an algorithm can be described that will compute successive approximations to $\sqrt{2}$.

What we see from these two examples is that the numbers π and $\sqrt{2}$ are "describable" by a finite set of ordinary symbols, namely the symbols appearing in the relevant algorithms. This brings up the question whether some real numbers are not "describable" in this sense. By an ordinary symbol we mean any symbol that a reader of this book would recognize: lowercase and uppercase Latin and Greek letters, digits, and other common symbols such as "space," "period," "left parenthesis," etc. We will call any such symbol a **letter**.

By a **word** we mean a finite sequence of letters. Examples of words are

good grade

Remember "space" is a letter.

which is a 10-letter word, and

$$\sup\{x \in \mathbf{R} : x^2 < 2\}$$

which is a 13-letter word. This book is also a word.

Proposition 13.35. *The set of words is countable.*

We call $x \in \mathbf{R}$ **describable** if there exist $m \in \mathbf{N}$ and an m-letter algorithm that computes the decimal expansion of x in the manner described above. Since there are only countably many words, there are only countably many algorithms. It follows that many real numbers are not describable. In fact, the set of nondescribable real numbers is uncountable.

In the earlier history of mathematics, it was thought that there was a chasm separating the "continuous" from the "discrete," or, if you like, **R** from **Z**. Gradually it became clear that all real numbers can be understood in terms of integers via decimals or suprema. But the chasm reappears in a more subtle way. While there exists a decimal representation for each real number, we now see that for most real numbers a decimal description cannot actually be written down.

Project 13.36. Which of the axioms for **R** are satisfied by the set of all describable numbers?

In technical language, the describable numbers form an ordered field.

Review Question. In what sense is the set **Z** of integers smaller than the set **R** of real numbers?

Weekly reminder: Reading mathematics is not like reading novels or history. You need to think slowly about every sentence. Usually, you will need to reread the same material later, often more than one rereading.

This is a short book. Its core material occupies about 140 pages. Yet it takes a semester for most students to master this material. In summary: read line by line, not page by page.

Chapter 14
Final Remarks

We have had several purposes in this book:

1. To teach you to *read mathematics* by encouraging you to read your own mathematics: to know the difference between an incorrect argument and a correct argument.
2. To teach you to *do mathematics*: to discover theorems and write down your own proofs, so that what you write down accurately reflects what you discovered and is free of mistakes. As time goes on, your style of writing mathematics will improve: watch how the writers of your textbooks write. Develop opinions about good and bad writing.
3. To teach you to *write mathematics* so that it is communicated accurately and clearly to another qualified reader.
4. To introduce you to the *axiomatic method*. This was explained in Chapter 1, but the point may not have been clear at the beginning. Please reread Chapter 1.
5. To teach you *induction*, one of the most fundamental tools.
6. To make you *understand the real numbers* and how the rational numbers are distributed in them.
7. To put the *whole mathematics curriculum* from Sesame Street through calculus in perspective.

Final Project. Discuss whether the following lines (from the poem *Little Gidding* in "Four Quartets" by T.S. Eliot) are relevant to the course you have just taken:

We shall not cease from exploration
And the end of all our exploring
Will be to arrive where we started
And know the place for the first time.

M. Beck and R. Geoghegan, *The Art of Proof: Basic Training for Deeper Mathematics*, Undergraduate Texts in Mathematics, DOI 10.1007/978-1-4419-7023-7_14,

Further Topics

Appendix A
Continuity and Uniform Continuity

First, it is neccessary to study the facts, to multiply the number of observations, and then later to search for formulas that connect them so as thus to discern the particular laws governing a certain class of phenomena. In general, it is not until after these particular laws have been established that one can expect to discover and articulate the more general laws that complete theories by bringing a multitude of apparently very diverse phenomena together under a single governing principle.
Augustin Louis Cauchy (1789–1857)

It is likely that your first calculus class included a discussion of continuity. Many students find the definition hard to understand, and in many calculus classes the fine details are skipped. One of the rewards of mastering the kind of mathematics discussed in this book is that items like the ε-δ definition of continuity are suddenly revealed as quite easy.

A.1 Continuity at a Point

If $a \in \mathbf{R}$ and $\delta \in \mathbf{R}_{>0}$, the **$\delta$-interval about** a is the open interval $(a-\delta, a+\delta)$. We think of it as a "neighborhood" of a.

In this chapter ε and δ will always denote positive real numbers, and for a given function f we will consider circumstances in which f maps the δ-interval about a into the ε-interval about $f(a)$.

The function $f : \mathbf{R} \to \mathbf{R}$ is **continuous at** a if for each $\varepsilon > 0$ there exists $\delta > 0$ such that f maps the δ-interval about a into the ε-interval about $f(a)$. Note that the number δ depends on both a and ε, but for now, a is fixed, so it is mainly important to think of δ as dependent on ε.

In Section A.3 below, the dependence on a as a varies will become important.

This is a subtle and important definition, so we will say it in a number of other ways:

(i) $\forall\, \varepsilon > 0\ \exists\, \delta > 0$ such that $f((a-\delta, a+\delta)) \subseteq (f(a)-\varepsilon, f(a)+\varepsilon)$.

(ii) $\forall\, \varepsilon > 0\ \exists\, \delta > 0$ such that $|x-a| < \delta \Rightarrow |f(x)-f(a)| < \varepsilon$.

Though you have seen (iv) in calculus, it has not been defined in this book. In fact, any of (i)–(iii) (they are all equivalent) defines (iv).

M. Beck and R. Geoghegan, *The Art of Proof: Basic Training for Deeper Mathematics*, Undergraduate Texts in Mathematics, DOI 10.1007/978-1-4419-7023-7_15,

(iii) $\forall \varepsilon > 0\, \exists \delta > 0$ such that when a number is distant $< \delta$ from a then its f-image is distant $< \varepsilon$ from $f(a)$.

(iv) $\lim_{x \to a} f(x) = f(a)$.

Two numbers in $\mathbf{R}$ are **α-close** if the distance from one to the other is less than α. The idea of continuity at a is that small intervals about a are mapped by f into small intervals about $f(a)$. But the word "small" has no objective meaning. So instead we say, "you tell us your idea of small (say, $\varepsilon > 0$) and we will guarantee you that there exists a (possibly much smaller) number $\delta > 0$ such that numbers δ-close to a are mapped by f to numbers ε-close to $f(a)$."

It took a long time in the history of mathematics to come up with this clever definition.

Example A.1. The function $f : \mathbf{R} \to \mathbf{R}$ given by $f(x) = x^2 + 1$ is continuous at $a = 3$ because

$$\lim_{x \to 3} f(x) = \lim_{x \to 3} x^2 + 1 = 10 = f(3).$$

(You should prove this.)

A.2 Continuity on a Subset of R

Recall that the open interval $(b, c) \subseteq \mathbf{R}$ is a set of the form $\{x : b < x < c\}$; the value $-\infty$ is permitted for b and ∞ for c. A subset of $\mathbf{R}$ is **open** if it is the union of open intervals. If $U \subseteq \mathbf{R}$ is open, one may as well think of U as a union of open intervals any two of which are disjoint, because the union of two open intervals that have a point in common is again an open interval (think about why this is so).

We need a new term in discussing functions. If $f : A \to B$ is a function and if $C \subseteq B$ we define the **preimage** of C to be

$$f^{-1}(C) := \{x \in A : f(x) \in C\}.$$

The preimage of C is also called the inverse image *of C.*
If C consists of just one number c, i.e., $C = \{c\}$, one usually writes $f^{-1}(c)$ rather than $f^{-1}(\{c\})$.

The function $f : \mathbf{R} \to \mathbf{R}$ is **continuous** if the preimage of every open set is an open set.

Proposition A.2. *f is continuous if and only if the preimage of every interval is an open set.*

Note that we are defining "continuous" as distinct from "continuous at a point." However, there is a nice relationship between the two definitions:

Proposition A.3. *f is continous if and only if f is continuous at every point $a \in \mathbf{R}$.*

Sometimes a function of interest is not defined on all of $\mathbf{R}$ but only on a subset $D \subseteq \mathbf{R}$,

An example is $f : \mathbf{R} - \{0\} \to \mathbf{R}$ given by $f(x) = \frac{1}{x}$.

so that we are dealing with $f : D \to \mathbf{R}$. Then we say that f is **continous at** $a \in D$ if

$$\forall \varepsilon > 0 \, \exists \delta > 0 \text{ such that } f((a-\delta, a+\delta) \cap D) \subseteq (f(a)-\varepsilon, f(a)+\varepsilon).$$

In other words, it is the same definition as before but now one considers only points of D. And similarly, we say that f is **continuous on** D if the preimage of every interval is the intersection of D with an open set.

Proposition A.4. *f is continuous on D if and only if*

$$\forall a \in D \, \forall \varepsilon > 0 \, \exists \delta > 0 \textit{ such that } f((a-\delta, a+\delta) \cap D) \subseteq (f(a)-\varepsilon, f(a)+\varepsilon).$$

In this Proposition δ depends on a as well as on ε.

A.3 Uniform Continuity

Consider the function $f : \mathbf{R} - \{0\} \to \mathbf{R}$ given by $f(x) = \frac{1}{x}$. If $a > 0$ this function is continuous at a. We now discuss why this is so:

Proposition A.5. *Let $f : \mathbf{R} - \{0\} \to \mathbf{R}$ be given by $f(x) = \frac{1}{x}$ and let $0 < \varepsilon < a$. The preimage of the open interval*

$$\left(\frac{1}{a} - \varepsilon, \frac{1}{a} + \varepsilon\right) \qquad \textit{is the open interval} \qquad \left(\frac{a}{1+a\varepsilon}, \frac{a}{1-a\varepsilon}\right).$$

A moment's calculation shows that a is not the midpoint of the interval $(\frac{a}{1+a\varepsilon}, \frac{a}{1-a\varepsilon})$. In fact, the number $a - \frac{a}{1+a\varepsilon}$ is smaller than the number $\frac{a}{1-a\varepsilon} - a$. So

$$f((a-\delta, a+\delta)) \subseteq (f(a)-\varepsilon, f(a)+\varepsilon)$$

only when $\delta \leq a - \frac{a}{1+a\varepsilon}$. Thus, for this given a and ε we have found the biggest possible δ. This δ depends on a as well as on ε. One might ask whether there is a smaller δ that works for all a (once ε is fixed). The answer is given in our next proposition.

Proposition A.6. *Let $\varepsilon > 0$ be given. There exists no number $\delta > 0$ such that for all $a > 0$ the function $f : \mathbf{R} - \{0\} \to \mathbf{R}$ given by $f(x) = \frac{1}{x}$ maps the interval $(a-\delta, a+\delta)$ into $(f(a)-\varepsilon, f(a)+\varepsilon)$.*

Note in this proposition the words "for all a." Once $\varepsilon > 0$ is fixed, there is a suitable δ for each a, but there is no number δ that will work for all a. This example suggests a new definition:

We say that $f : D \to \mathbf{R}$ is **uniformly continuous on** $D \subseteq \mathbf{R}$ if

$$\forall \varepsilon > 0 \, \exists \delta > 0 \text{ such that } \forall a \in D, \; f((a-\delta, a+\delta) \cap D) \subseteq (f(a)-\varepsilon, f(a)+\varepsilon).$$

Here, δ might depend on ε but does not depend on a.

Example A.7. We claim that the function $f : \mathbf{R} \to \mathbf{R}$ given by $f(x) = x^2 + 1$ is uniformly continuous on $[1,4]$ but not on $\mathbf{R}$.

The first part of our claim says

$$\forall \varepsilon > 0 \ \exists \delta > 0 \text{ such that } \forall a \in [1,4],\ f((a-\delta, a+\delta)) \subseteq (f(a) - \varepsilon, f(a) + \varepsilon).$$

To prove this, given $\varepsilon > 0$, there exists (by Proposition 10.4) a positive number $\delta \leq \min\left\{1, \frac{\varepsilon}{9}\right\}$. For each $a \in [1,4]$ we have the two inequalities

$$\begin{aligned}(a+\delta)^2 + 1 &= a^2 + 1 + (2a+\delta)\delta \\ &\leq a^2 + 1 + (8+\delta)\delta && \text{(because } a \leq 4) \\ &\leq a^2 + 1 + 9\delta && \text{(because } \delta \leq 1) \\ &\leq a^2 + 1 + \varepsilon && \text{(because } \delta \leq \tfrac{\varepsilon}{9}) \\ &= f(a) + \varepsilon\end{aligned}$$

and

$$\begin{aligned}(a-\delta)^2 + 1 &= a^2 + 1 + (-2a+\delta)\delta \\ &\geq a^2 + 1 + (-8+\delta)\delta && \text{(because } a \leq 4) \\ &\geq a^2 + 1 - 8\delta \\ &\geq a^2 + 1 - \varepsilon && \text{(because } 8\delta < 9\delta \leq \varepsilon) \\ &= f(a) - \varepsilon.\end{aligned}$$

Why is it important to note that f is increasing on $[1,4]$?

Since f is an increasing function on $[1,4]$, this implies

$$f((a-\delta, a+\delta)) = \left((a-\delta)^2 + 1, (a+\delta)^2 + 1\right) \subseteq \left(a^2 + 1 - \varepsilon, a^2 + 1 + \varepsilon\right).$$

Our choice of δ depended on ε but not on $a \in [1,4]$.

The second part of our claim says

$$\begin{aligned}&\exists \varepsilon > 0 \ \text{ such that } \ \forall \delta > 0 \ \exists a \in \mathbf{R} \ \text{ such that} \\ &f((a-\delta, a+\delta)) \not\subseteq (f(a) - \varepsilon, f(a) + \varepsilon).\end{aligned}$$

For this it suffices to prove

$$\exists \varepsilon > 0 \ \text{ such that } \ \forall \delta > 0 \ \exists a \in \mathbf{R} \text{ such that } \ f(a+\delta) > f(a) + \varepsilon.$$

Let $\varepsilon = 1$. Then given any $\delta > 0$, let $a = \frac{1}{2\delta}$. Then

$$(a+\delta)^2 + 1 = a^2 + 2a\delta + \delta^2 + 1 > a^2 + 2a\delta + 1 = a^2 + 2 = a^2 + 1 + \varepsilon,$$

in other words, $f(a+\delta) > f(a)+\varepsilon$. Thus f is not uniformly continuous on $[0,\infty)$ (where f is increasing), hence not on **R**. □

In analysis the distinction between continuity and uniform continuity of a function with domain $D \subseteq \mathbf{R}$ is often important. You will prove the following famous theorem in your first analysis course.

Theorem A.8. *If D is a closed interval in* **R** *and if $f : D \to \mathbf{R}$ is continuous on D then f is uniformly continuous on D.*

The proof of this involves a topological idea called "compactness," which will not be discussed here.

Appendix B
Public-Key Cryptography

The purpose of computation is insight, not numbers.
Richard Hamming (1915–1998)

In this chapter, we will explore some computational aspects of modular arithmetic, which we studied in Chapter 6. We will be concerned about how to compute certain numbers in a most efficient way. There is a whole field at work here, of which we barely scratch the surface, called *computational complexity*. For example, in Section 6.4 we introduced the concept of greatest common divisor (gcd). You might wonder how quickly one could compute the gcd of two, say, 1000-digit integers. As another example, we now discuss how to quickly compute a^b modulo c for given positive integers a, b, and c.

B.1 Repeated Squaring

What are the last two digits of 58^{231}? Mathematically, we are asking which integer between 0 and 99 is congruent to 58^{231} modulo 100. There is a long way to compute this and a short one:

First approach. Compute $58 \cdot 58$, reduce the result mod 100, multiply by 58, reduce, multiply, reduce, multiply...

Second approach. Compute the binary expansion of 231:

$$231 = 2^7 + 2^6 + 2^5 + 2^2 + 2^1 + 2^0.$$

Now square repeatedly, at each step reducing mod 100:

M. Beck and R. Geoghegan, *The Art of Proof: Basic Training for Deeper Mathematics*, Undergraduate Texts in Mathematics, DOI 10.1007/978-1-4419-7023-7_16,

$$\begin{aligned}
58^2 &= 3364 \equiv 64 \pmod{100} \\
58^4 &= \left(58^2\right)^2 \equiv 64^2 = 4096 \equiv 96 \pmod{100} \\
58^8 &= \left(58^4\right)^2 \equiv 96^2 \equiv 16 \pmod{100} \\
58^{16} &\equiv 16^2 \equiv 56 \pmod{100} \\
58^{32} &\equiv 56^2 \equiv 36 \pmod{100} \\
58^{64} &\equiv 36^2 \equiv 96 \pmod{100} \\
58^{128} &\equiv 96^2 \equiv 16 \pmod{100}.
\end{aligned}$$

Now we can piece everything together:

$$58^{231} = 58^{128+64+32+4+2+1} \equiv 16 \cdot 96 \cdot 36 \cdot 96 \cdot 64 \cdot 58 \equiv 92 \pmod{100}.$$

The process in our second approach is called *repeated squaring*.

Project B.1. Compare the running times (number of steps needed) of these two approaches.

The process of repeated squaring is extremely useful in computations. It will implicitly appear throughout the remainder of this chapter.

Hint: use Corollary 6.36.

Proposition B.2. *If p is a prime that does not divide $k \in \mathbf{Z}$, then there exists an integer k^{-1} such that*

$$k^{-1}k \equiv 1 \pmod{p}.$$

We use the notation $a \pmod{n}$ to denote the least nonnegative integer congruent to a modulo n.

If you have followed the hint to prove Proposition B.2, you will have found out that $k^{-1} \equiv k^{p-2} \pmod{p}$. In practice, k and p might be large, and you will use repeated squaring to compute $k^{p-2} \pmod{p}$.

B.2 Diffie–Hellman Key Exchange

You (Y) and your friend (F) would like to devise a method of exchanging secret messages. Being mathematicians, you have long agreed on a way to encode letters into numbers (for example, using the ASCII system), so we may as well assume that the secret messages are positive integers. In practice, these will be large numbers, but for the purpose of computing some explicit examples, we assume that the messages are broken into pieces such that each individual message is a 2-digit number. Y and F would like to come up with a key k to encode (and later decode) a given message m.

Here is one simple scheme:

In practice p will be much larger.

- Y and F agree on a 3-digit prime p and a key k that is not divisible by p.

- Y encrypts the message m by computing $km \pmod p$ and sends the result to F.

Proposition B.2 asserts the existence of k^{-1} (which can be computed using repeated squaring!), and this number works as the decryption key:

- F takes the incoming message and multiplies it by k^{-1}, yielding

$$k^{-1}(km) = \left(k^{-1}k\right) m \equiv m \pmod p .$$

Since $m < p$, this returns m, and so F got Y's message.

Example B.3. Y and F agree on $p = 113$ and $k = 34$. One computes (just do it!) that $k^{-1} \equiv 10 \pmod{113}$. Y wants to transmit the message $m = 42$ to F, so Y encrypts

$$km = 42 \cdot 34 \equiv 72 \pmod{113}$$

and sends the number 72 to F. F knows how to decrypt this number:

$$k^{-1} \cdot 72 = 10 \cdot 72 \equiv 42 \pmod{113} .$$

Project B.4. Use the above scheme to encrypt some simple messages (for example, using the numbers 1 through 26 for the letters of the alphabet—send one letter at a time). Get one of your friends to decrypt your messages.

Here is the catch: Y and F live far away from each other. They can call or email each other, but they have to assume that their communications are not secure. So they have to create p and the encryption key k in such a way that the process is public, yet only they know k in the end. This procedure is called *public-key exchange*. The first viable public-key exchange was discovered by Whitfield Diffie and Martin Hellman in 1976. Here is how it works.

The Diffie–Hellman key exchange is still used today, for example, in some ssh protocols.

- Y and F (publicly) agree on a prime p and a positive integer $a < p$.
- Y secretly thinks of a positive integer y, computes $a^y \pmod p$ (using repeated squaring), and sends the result to F.
- F secretly thinks of a positive integer f, computes $a^f \pmod p$ (using repeated squaring), and sends the result to Y.

The encryption key k that Y and F can now use is

$$k \equiv a^{yf} \pmod p .$$

Both Y and F can compute this number because $k \equiv (a^y)^f = \left(a^f\right)^y \pmod p$. (Once more they will use repeated squaring in the actual computation.)

Example B.5. We work again with $p = 113$ to keep this example simple. Y and F agree to use $a = 22$. Y comes up with $y = 69$, computes $22^{69} \equiv 88 \pmod{113}$, and

Note once more how important repeated squaring is in our various computations.

sends the number 88 to F. F comes up with $f = 38$, computes $22^{38} \equiv 14 \pmod{113}$, and sends the number 14 to Y. Thus the encryption key that Y and F will use is $k = 22^{69\cdot 38} \equiv 8 \pmod{113}$.

Project B.6. Work with your friend through another example of a Diffie–Hellman key.

How could a third party come across k? Since Y and F's communication is not secure, we may assume that the third party knows p, a, $a^y \pmod{p}$, and $a^f \pmod{p}$. To compute k, the third party would need to know either y or f, but those are secret. (Note that Y does not have to know f, and F does not have to know y, in order to compute the encryption key k.)

The problem of computing y from a and $a^y \pmod{p}$ is known as the discrete log problem.

The Diffie–Hellman key exchange is based on the "fact" that it is computationally hard to compute y when we know only a and $a^y \pmod{p}$. We used quotation marks because this "fact" is merely the status quo: nobody has been able to prove that there is no quick way of computing y from a and $a^y \pmod{p}$. Perhaps someday someone will prove that the solution is computationally feasible, or someone will prove the opposite.

Project B.7. Consider this alternative to the Diffie–Hellman scheme:

- Y and F (publicly) agree on a prime p and a positive integer $a < p$.
- Y secretly thinks of a positive integer y, computes $ay \pmod{p}$, and sends the result to F.
- F secretly thinks of a positive integer f, computes $af \pmod{p}$, and sends the result to Y.

The encryption key k that Y and F can now use is $k \equiv ayf \pmod{p}$, and as before, both Y and F can easily compute k. Discuss the security of this public-key exchange.

RSA is named after Ron Rivest, Adi Shamir, and Leonard Adleman.

Project B.8. Do a research project on RSA public-key encryption. Describe how and why RSA works, and discuss the advantages and disadvantages compared to Diffie–Hellman.

Appendix C
Complex Numbers

The imaginary number is a fine and wonderful resource of the human spirit, almost an amphibian between being and not being.
Gottfried Leibniz (1646–1716)

One deficiency of the real numbers is that the equation $x^2 = -1$ has no solution $x \in \mathbf{R}$ (Corollary 11.23). In this chapter, we will extend $\mathbf{R}$ to overcome this deficiency; the price that we will have to pay is that this extension does not have a useful ordering relation.

C.1 Definition and Algebraic Properties

A **complex number** is an ordered pair of real numbers. The set of all complex numbers is denoted by $\mathbf{C} := \{(x,y) : x,y \in \mathbf{R}\}$. $\mathbf{C}$ is equipped with the **addition**

$$(x,y) + (a,b) := (x+a, y+b)$$

and the **multiplication**

$$(x,y) \cdot (a,b) := (xa - yb, xb + ya).$$

Just as we embedded $\mathbf{Z}$ in $\mathbf{R}$, we embed $\mathbf{R}$ in $\mathbf{C}$ by the injective function $e : \mathbf{R} \to \mathbf{C}$, $e(r) = (r,0)$. Identifying r with $e(r)$, we will write $\mathbf{R} \subseteq \mathbf{C}$ from now on.

You should convince yourself that e preserves addition and multiplication.

Proposition C.1. *For all* $(a,b),(c,d),(e,f) \in \mathbf{C}$*:*

(i) $(a,b) + (c,d) = (c,d) + (a,b)$.

(ii) $((a,b) + (c,d)) + (e,f) = (a,b) + ((c,d) + (e,f))$.

(iii) $(a,b) \cdot ((c,d) + (e,f)) = (a,b) \cdot (c,d) + (a,b) \cdot (e,f)$.

M. Beck and R. Geoghegan, *The Art of Proof: Basic Training for Deeper Mathematics*, Undergraduate Texts in Mathematics, DOI 10.1007/978-1-4419-7023-7_17,

(iv) $(a,b)\cdot(c,d) = (c,d)\cdot(a,b)$.

(v) $((a,b)\cdot(c,d))\cdot(e,f) = (a,b)\cdot((c,d)\cdot(e,f))$.

Proposition C.2. *There exists a complex number* 0 *such that for all* $z \in \mathbf{C}$, $z+0=z$.

Proposition C.3. *There exists a complex number* 1 *with* $1 \neq 0$ *such that for all* $z \in \mathbf{C}$, $z\cdot 1 = z$.

Proposition C.4. *For each* $z \in \mathbf{C}$, *there exists a complex number, denoted by* $-z \in \mathbf{C}$, *such that* $z+(-z)=0$.

Thus **C** *satisfies Axioms 8.1–8.5.*

Proposition C.5. *For each* $z \in \mathbf{C}-\{0\}$, *there exists a complex number, denoted by* z^{-1}, *such that* $z\cdot z^{-1} = 1$.

The last two propositions allow us to define **subtraction** and **division** of two complex numbers, just as in the real case.

The definition of our multiplication implies the innocent-looking statement

$$(0,1)\cdot(0,1) = (-1,0). \tag{C.1}$$

This equation together with the fact that

$$(a,0)\cdot(x,y) = (ax,ay)$$

leads to an alternative notation for complex numbers—the notation that is always used—as we now explain. We can write

$$(x,y) = (x,0)+(0,y) = (x,0)\cdot(1,0)+(y,0)\cdot(0,1).$$

If we think—in the spirit of our remark on the embedding of **R** in **C**—of $(x,0)$ and $(y,0)$ as the real numbers x and y, then we can write any complex number (x,y) as a linear combination of $(1,0)$ and $(0,1)$, with the real coefficients x and y. Now, $(1,0)$ can be thought of as the real number 1. So if we give $(0,1)$ a special name, the traditional choice is i, then the complex number that we have been writing as $z=(x,y)$ can be written as $x\cdot 1 + y\cdot i$, or in short,

$$z = x+iy.$$

The name has historical origins: people thought of complex numbers as unreal, imagined.

The number x is called the **real part** and y the **imaginary part** of the complex number $x+iy$, often denoted by $\operatorname{Re}(x+iy)=x$ and $\operatorname{Im}(x+iy)=y$.

The equation (C.1) now reads

$$i^2 = -1,$$

so that $i \in \mathbf{C}$ is a solution to the equation $z^2=-1$.

One can say much more: every polynomial equation has a solution in $\mathbf{C}$. This fact, the *Fundamental Theorem of Algebra*, is too difficult for us to prove here. But you will see a proof if you take a course in complex analysis (which we strongly recommend). While we gained solutions to previously unsolvable equations by extending $\mathbf{R}$ to $\mathbf{C}$, this came at a price:

With Proposition 6.20, the Fundamental Theorem of Algebra implies that every polynomial of degree d has d roots in $\mathbf{C}$.

Project C.6. Discuss the sense in which $\mathbf{C}$ does not satisfy Axiom 8.26.

C.2 Geometric Properties

Although we just introduced a new way of writing complex numbers, we briefly return to the (x,y)-notation. It suggests that one can think of a complex number as a two-dimensional real vector. When plotting these vectors in the plane $\mathbf{R}^2$, we will call the x-axis the **real axis** and the y-axis the **imaginary axis**. The addition that we defined for complex numbers resembles vector addition.

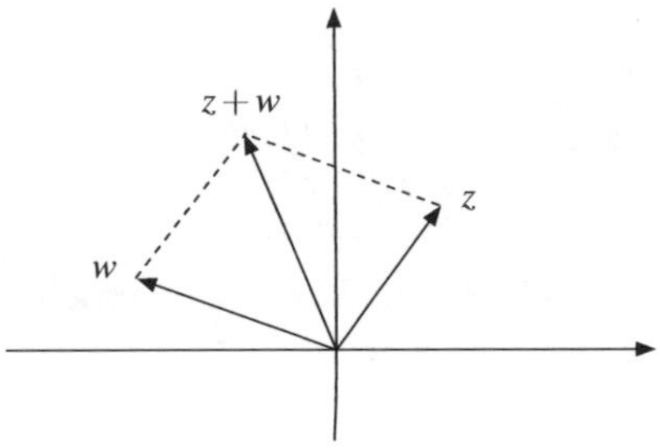

Fig. C.1 Addition of complex numbers.

Any vector in $\mathbf{R}^2$ is defined by its two coordinates. On the other hand, it is also determined by its length and the angle it encloses with, say, the positive real axis; we now define these concepts thoroughly. The **absolute value** (also called the **modulus**) of $x+iy$ is

$$r := |x+iy| = \sqrt{x^2+y^2},$$

and an **argument** of $x+iy$ is a number ϕ such that

Here we need to assume high-school trigonometry.

$$x = r\cos\phi \qquad \text{and} \qquad y = r\sin\phi .$$

This means, naturally, that any complex number has many arguments; any two of them differ by a multiple of 2π.

The absolute value of the difference of two vectors has a geometric interpretation: it is the *distance* between the (end points of the) two vectors (see Figure C.2). It is very useful to keep this geometric interpretation in mind when thinking about the absolute value of the difference of two complex numbers.

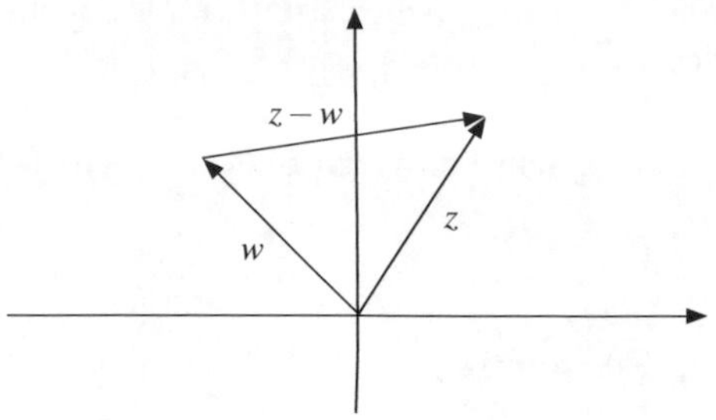

Fig. C.2 Geometry behind the distance of two complex numbers.

The first hint that absolute value and argument of a complex number are useful concepts is the fact that they allow us to give a geometric interpretation for the multiplication of two complex numbers.

Proposition C.7. *If* $x_1 + iy_1 \in \mathbf{C}$ *has absolute value* r_1 *and argument* ϕ_1, *and* $x_2 + iy_2 \in \mathbf{C}$ *has absolute value* r_2 *and argument* ϕ_2, *then the product* $(x_1 + iy_1)(x_2 + iy_2)$ *has absolute value* $r_1 r_2$ *and argument (one among many)* $\phi_1 + \phi_2$.

You should convince yourself that there is no problem with the fact that there are many possible arguments for complex numbers, since both cosine and sine are periodic functions with period 2π.

Geometrically, we are multiplying the lengths of the two vectors representing our two complex numbers, and adding their angles measured with respect to the positive real axis.

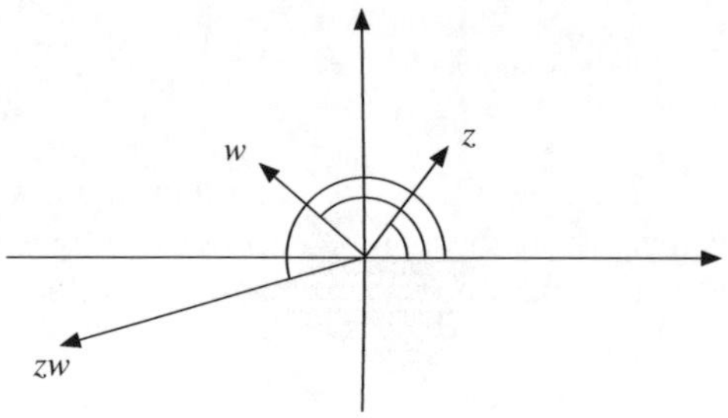

Fig. C.3 Multiplication of complex numbers.

The notation $e^{i\phi} := \cos\phi + i\sin\phi$ is handy: with this notation, the sentence "The complex number $x + iy$ has absolute value r and argument ϕ" now becomes the equation

$$x + iy = re^{i\phi}.$$

The left-hand side is often called the **rectangular form**, the right-hand side the **polar form** of this complex number.

At this point, this exponential notation is indeed purely a notation, but it has an intimate connection to the complex exponential function, of which you will see a glimpse at the end of this chapter. For now we motivate our use of this notation by the following proposition.

Proposition C.8. *For all* $\phi, \psi \in \mathbf{R}$,

$$\begin{aligned} e^{i\phi}\, e^{i\psi} &= e^{i(\phi+\psi)}\,, \\ 1/e^{i\phi} &= e^{-i\phi}\,, \\ e^{i(\phi+2\pi)} &= e^{i\phi}\,, \\ \left|e^{i\phi}\right| &= 1\,. \end{aligned}$$

Proposition C.9. *For all* $z \in \mathbf{C}$, $x, y \in \mathbf{R}$,

(i) $-|z| \le \operatorname{Re} z \le |z|$.

(ii) $-|z| \le \operatorname{Im} z \le |z|$.

(iii) $|x+iy|^2 = (x+iy)(x-iy)$.

The last equation of this proposition is one of many reasons to give the process of passing from $x+iy$ to $x-iy$ a special name: $x-iy$ is called the **(complex) conjugate** of $x+iy$. We denote the conjugate by

$$\overline{x+iy} := x - iy\,.$$

Geometrically, conjugating z means reflecting the vector corresponding to z in the real axis—think of the real axis as a mirror. Here are some basic properties of the conjugate.

Proposition C.10. *For all* $z, w \in \mathbf{C}$, $\phi \in \mathbf{R}$,

(i) $\overline{z \pm w} = \overline{z} \pm \overline{w}$,

(ii) $\overline{z \cdot w} = \overline{z} \cdot \overline{w}$,

(iii) $\overline{z/w} = \overline{z}/\overline{w}$,

(iv) $\overline{(\overline{z})} = z$,

(v) $|\overline{z}| = |z|$,

(vi) $|z|^2 = z\overline{z}$,

(vii) $\operatorname{Re} z = \frac{1}{2}\left(z + \overline{z}\right)$,

(viii) $\operatorname{Im} z = \frac{1}{2i}\left(z - \overline{z}\right)$,

(ix) $\overline{e^{i\phi}} = e^{-i\phi}$.

Here is the complex counterpart to Proposition 10.10(iv), the *triangle inequality*.

If you draw a picture you will see the reason behind the name "triangle inequality."

Proposition C.11. *For all* $z, w \in \mathbf{C}$,

$$|z+w| \le |z| + |w| .$$

For every z this series converges to a complex number in a sense analogous to convergence of real series as defined in Section 12.1.

We cannot write a chapter on complex numbers without mentioning the **(complex) exponential function**

$$\exp(z) := \sum_{j=0}^{\infty} \frac{z^j}{j!} .$$

Project C.12. Write out the Taylor series for $\cos\theta$ and $\sin\theta$, which you have seen in calculus. Combine them to get a complex Taylor series for $\cos\theta + i\sin\theta$. Compare the result with what you get in the above series for $\exp(z)$ when $z = i\theta$.

Appendix D
Groups and Graphs

"And what is the use of a book," thought Alice, "without pictures or conversations?"
Lewis Carroll (*Alice in Wonderland*)

D.1 Groups

A **group** is a set G equipped with a binary operation, $\cdot$, and a special element, $1 \in G$, satisfying the following axioms:

(i) For all $g, h, k \in G$, $(g \cdot h) \cdot k = g \cdot (h \cdot k)$.

(ii) For each $g \in G$, $g \cdot 1 = g$.

(iii) For each $g \in G$ there exists $g^{-1} \in G$ such that $g \cdot g^{-1} = 1$.

The binary operation $\cdot$ is usually described as **multiplication** or simply as the **group operation**; 1 is called the **identity element** or just the **identity**; g^{-1} is called the **inverse** of g. As with numbers, it is common to write gh for $g \cdot h$, and we will often do that here.

Proposition D.1. *Each $g \in G$ has only one inverse; in other words, if h_1 and h_2 are inverses for g, then $h_1 = h_2$. The element 1 is also unique in this sense.*

In the above definition, 1 is presented as a right identity and g^{-1} appears as a right inverse. However, one easily proves that they also have these properties from the left:

Proposition D.2. *For all $g \in G$, $1 \cdot g = g$ and $g^{-1} \cdot g = 1$.*

Example D.3. Any set having just one member becomes a group in a rather obvious way: We may as well name the single member 1. The multiplication is defined by $1 \cdot 1 = 1$. Then $(1 \cdot 1) \cdot 1 = 1 \cdot 1 = 1$ and, similarly, $1 \cdot (1 \cdot 1) = 1$, so the multiplication

M. Beck and R. Geoghegan, *The Art of Proof: Basic Training for Deeper Mathematics*, Undergraduate Texts in Mathematics, DOI 10.1007/978-1-4419-7023-7_18,

is associative. Clearly $1^{-1} = 1$. This group is called the **trivial group**. (There is some loose language here: any group having one member is a trivial group in this sense. One usually calls any such group "the" trivial group.)

Example D.4. We have already seen some groups in this book. G could be $\mathbf{Z}$ with $+$ as the group operation and 0 as the identity element; in that case $n^{-1} = -n$. Or G could be the set of positive rational numbers with the usual multiplication as the group operation and 1 as the identity element; then $(\frac{m}{n})^{-1} = \frac{n}{m}$.

Abelian groups are named after Niels Henrik Abel (1802–1829); a more practical name would have been commutative group.

These examples have the feature that the multiplication is commutative, i.e., $gh = hg$ for all $g, h \in G$. A group satisfying that additional property is **Abelian**.

Example D.5. If A is a set, a bijection $A \to A$ is called a **permutation** of A. Consider the set $[n]$ whose members are the positive integers $1, 2, \ldots, n$. Let S_n denote the set of all permutations of $[n]$. Define a multiplication on S_n by $f \cdot g = f \circ g$; in other words, composition of functions is the group operation. You have probably noticed already in Chapter 9 that this operation is associative. The identity map of $[n]$ plays the role of 1. Since the elements of S_n are bijections, each element has an inverse. In this way S_n is a group, called the nth **symmetric group**. S_n has $n!$ members.

In general, for functions, $f \circ (g \circ h) = (f \circ g) \circ h$ whenever composition makes sense.

Proposition D.6. *S_2 is an Abelian group. When $n > 2$, the group S_n is not Abelian.*

SL stands for special linear; "linear" because these matrices can be thought of as linear automorphisms of a 2-dimensional real vector space, and "special" because their determinant is 1. $PSL_2(\mathbf{Z})$ is called the modular group and plays an important role in number theory and hyperbolic geometry.

Example D.7. Let $SL_2(\mathbf{Z})$ denote the set of all 2×2 matrices with integer entries and determinant 1. The group operation is matrix multiplication, which is associative (you can easily check this). The identity element is the matrix $I = \left[\begin{smallmatrix} 1 & 0 \\ 0 & 1 \end{smallmatrix}\right]$. The inverse of the matrix $\left[\begin{smallmatrix} a & b \\ c & d \end{smallmatrix}\right] \in SL_2(\mathbf{Z})$ is $\left[\begin{smallmatrix} d & -b \\ -c & a \end{smallmatrix}\right]$.

Example D.8. Place the following equivalence relation on the set $SL_2(\mathbf{Z})$: Matrices A and B are equivalent if $B = \pm A$. Let $PSL_2(\mathbf{Z})$ denote the set of equivalence classes; the equivalence class $\{A, -A\}$ is denoted by $[A]$. In a natural way $PSL_2(\mathbf{Z})$ becomes a group: the multiplication is defined by $[A] \cdot [B] = [AB]$. The identity element is $[I]$, and $[A]^{-1} = [A^{-1}]$. (One must check that multiplication in $PSL_2(\mathbf{Z})$ is well defined, independent of which representatives are chosen for $[A]$ and $[B]$.)

D.2 Subgroups

Why does this force 1 to lie in H?

A **subgroup** of the group G is a subset H that is closed under multiplication and closed under inversion. In other words, it is a group in its own right with respect to the multiplication defined on G.

Example D.9. The even integers form a subgroup of $\mathbf{Z}$ (with respect to $+$).

Example D.10. The dyadic rational numbers, i.e., positive fractions that in lowest terms have powers of 2 as their denominators, form a subgroup of the positive rationals (with respect to $\cdot$).

Example D.11. The set $\{\pm I\}$ is a two-element subgroup of $\mathrm{SL}_2(\mathbf{Z})$.

Example D.12. If G is a group with identity element 1 then the set $\{1\}$ is a subgroup of G; in other words, the trivial group is a subgroup of any group.

A permutation f of $[n]$ is an **even permutation** if the number of pairs of members $(j,k) \in [n] \times [n]$ such that $j < k$ and $f(j) > f(k)$ is even. If, on the other hand, this number is odd, then f is an **odd permutation**.

Proposition D.13. *The even permutations constitute a subgroup of S_n. This subgroup contains $\frac{n!}{2}$ members.*

This subgroup of S_n is called the nth **alternating group** and is denoted by A_n.

Proposition D.14. *Fix a positive integer N and let*

$$S := \left\{ \left[\begin{smallmatrix} a & b \\ c & d \end{smallmatrix}\right] \in \mathrm{SL}_2(\mathbf{Z}) : a,d \equiv 1 \,(\bmod N),\, b,c \equiv 0 \,(\bmod N) \right\}.$$

Then S is a subgroup of $\mathrm{SL}_2(\mathbf{Z})$.

S is an example of a congruence subgroup, *an important concept in number theory.*

D.3 Symmetries

The term *group* abbreviates the more descriptive term *group of symmetries*. To illustrate this we first consider a regular pentagon drawn in the plane; *regular* means that all five sides have the same length (from which it follows that all five angles are equal). This pentagon encloses a bounded region in the plane. Imagine that the plane is made of stiff plastic, and that you have cut along the sides of the pentagon to get a five-sided flat plate. A "symmetry" of this plate results from picking up the plate and placing it back down exactly where it was before, but with individual points perhaps occupying new positions (see Figure D.1). For example, you might have rotated the plate through 72 degrees ($=\frac{2\pi}{5}$ radians) about the plate's center point. Then each of the five vertices will have moved counterclockwise to occupy the position previously occupied by a neighboring vertex. Or you might have turned the plate upside down, returning it to exactly where it was before. We call the first of these a *rotation* and the second a *reflection*. There are five rotations and, following any of them by a reflection, there are five other symmetries that are not rotations. This set of symmetries is a group in the following sense: the multiplication is composition (we think of each symmetry as a bijection from the plate to itself). The result of doing two symmetries

The plate is rigid and so we consider only permutations that respect this rigidity; in mathematical terms, the distance between any two points of the plate must be preserved by each symmetry.

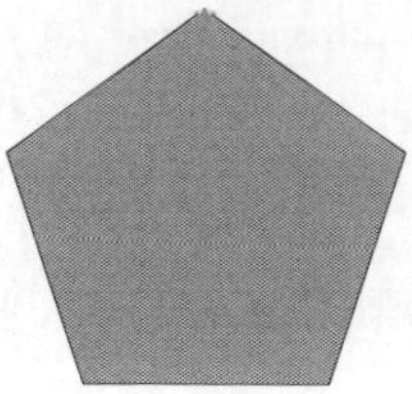

Fig. D.1 Pentagonal plate.

successively is another symmetry. The "do nothing" symmetry is the identity element and the "undo what you just did" symmetry is the inverse (of the symmetry you just performed).

Project D.15. Convince yourself that although it may seem that you can describe many more than ten (rigid) symmetries of this pentagon, only ten are different from one another.

Instead of considering the symmetries of our pentagonal plate, consider the symmetries of the slightly more complicated picture illustrated in Figure D.2. This object is

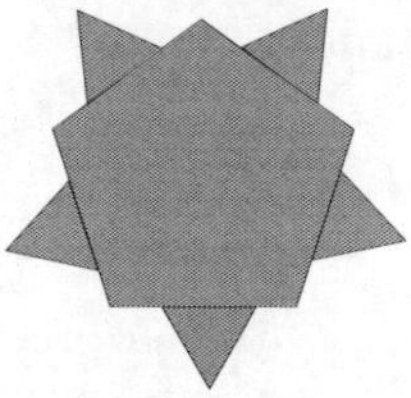

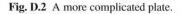

Fig. D.2 A more complicated plate.

obtained by gluing an equilateral triangle to each of the five sides of the pentagon as shown in the picture. Again, we regard this as a rigid plate. A moment's thought will convince you that the group of symmetries of this new figure is the same as the group of symmetries of the pentagonal plate.

This illustrates an important, if rather abstract, idea: mathematics distinguishes between a group of symmetries and the particular object that "realizes" that group of symmetries. As we have just seen, different objects may have the same group of symmetries. It takes a certain experience with abstract ideas to be comfortable with this. (Try explaining it to someone not trained in mathematics, as we have tried on occasion.) But focusing on this distinction marked a moment of real progress in the history of mathematics.

The meaning of the word "symmetry" depends on context. There is a sense in which every group is a group of symmetries of some mathematical object. We give an indication of this when we discuss Cayley graphs in Section D.6.

D.4 Finitely Generated Groups

When G is a group and $S \subseteq G$, we write S^{-1} for the set of inverses of members of S.

A group G is **finitely generated** if there exists a finite subset $S \subseteq G$ such that every element of G can be written as the product (under the group operation) of finitely many members of G each of which belongs to the set $S \cup S^{-1}$. Such a set S is a **set of generators** for G, and we say that S **generates** G.

If the group G is finite, for example the group of symmetries of the pentagonal plate, then it is immediate that G is finitely generated, but there may still be interest in choosing a set of generators smaller than the whole set G.

Example D.16. Let G be the 10-member group of symmetries of the pentagonal plate described in Section D.1. Let a be the rotation through 72 degrees, and let b be a reflection. Then $\{a,b\}$ is a set of generators for G.

Example D.17. The group $\mathbf{Z}$ with the operation $+$ is generated by the single element 1.

Example D.18. The group $\mathrm{SL}_2(\mathbf{Z})$ is generated by $A := \left[\begin{smallmatrix} 0 & 1 \\ -1 & 0 \end{smallmatrix}\right]$ and $B := \left[\begin{smallmatrix} 0 & 1 \\ -1 & 1 \end{smallmatrix}\right]$.

It is not obvious that A and B generate $\mathrm{SL}_2(\mathbf{Z})$.

Proposition D.19. *The multiplicative group of positive rational numbers is not finitely generated.*

Project D.20. What is the smallest possible number of generators of S_n?

Proposition D.21. *Every finitely generated group is countable.*

D.5 Graphs

A **graph** Γ consists of a nonempty countable set V of **vertices** and a set E of **edges**, where E is a collection of 2-element subsets of V. If $e = \{v_1, v_2\}$, we call the vertices v_1 and v_2 the **endpoints** of the edge e. We also say that v_1 and v_2 **span** e.

If you check the definition of "graph" in books you will find a number of variations; ours is restrictive but is enough for what we do here.

Consider a wire-frame model of the graph Γ in 3-dimensional space: each edge is represented by a rigid piece of wire of length ≥ 1 joining its two endpoints, and two different wire segments either are disjoint from one another or meet in one common endpoint. It is a theorem (which we will assume) that every graph can be modeled in this way.

This use of the word "graph" has no connection with "graph of a function."

The graph Γ is **connected** if two bugs sitting at different vertices can crawl along the wire frame until they meet. (This is not the formal definition of "connected"; the idea is that there should be at least one path joining any two vertices; the precise definition is too long to give here.)

In the next section we describe how to exhibit a group as a group of symmetries of a specially constructed connected graph.

D.6 Cayley Graphs

Cayley graphs are named in honor of Arthur Cayley (1821–1895).

Let G be a finitely generated group. Fix a finite generating set S; we will always assume that $1 \notin S$. The **Cayley graph** $\Gamma(G,S)$ has the members of G as its vertices; the elements $g,h \in G$ span an edge if and only if $h = gs$ for some $s \in S$. Our wording implies that even if both s and s^{-1} lie in S there is only one edge whose endpoints are g and gs.

Project D.22. Why did we exclude $1 \in S$?

Proposition D.23. *The Cayley graph $\Gamma(G,S)$ is a connected graph.*

The groups in Examples D.24 and D.25 are infinite, so we can only indicate a small piece of the Cayley graph, but enough to give an idea of the rest.

Example D.24. If G is $\mathbf{Z}$ with $+$ as the group operation and $S = \{1\}$, then for each $n \in \mathbf{Z}$ there is an edge joining n to $(n+1)$, and this rule accounts for all the edges. So the Cayley graph looks like Figure D.3.

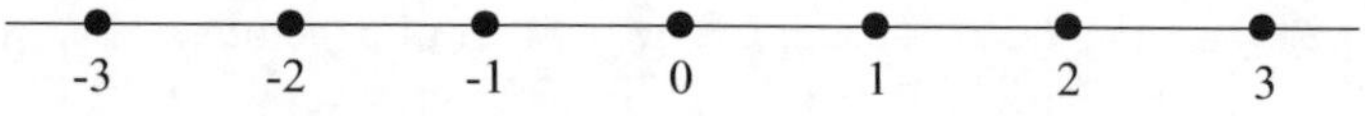

Fig. D.3 Part of the Cayley graph of $\mathbf{Z}$.

Example D.25. If G is $\mathbf{Z} \times \mathbf{Z}$ with the operation $+$ defined by

$$(a,b) + (c,d) := (a+c, b+d)$$

and $S = \{(1,0),(0,1)\}$ then for each $(m,n) \in \mathbf{Z} \times \mathbf{Z}$ there is an edge joining it to $(m+1,n)$ and another edge joining it to $(m,n+1)$, and this rule accounts for all the edges, so the Cayley graph looks like Figure D.4.

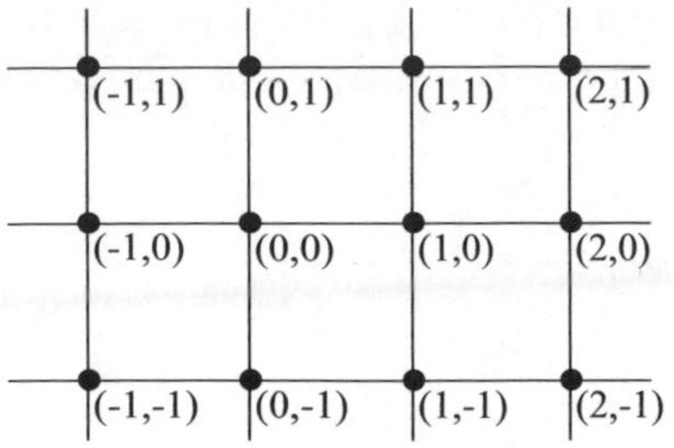

Fig. D.4 Part of the Cayley graph of $\mathbf{Z} \times \mathbf{Z}$.

Here is a finite example:

Project D.26. It can be shown that the group A_5 is generated by two permutations s and t given as follows:

- $s(1) = 4,\ s(2) = 1,\ s(3) = 5,\ s(4) = 3,\ s(5) = 2,$
- $t(1) = 2,\ t(2) = 1,\ t(3) = 4,\ t(4) = 3,\ t(5) = 5.$

These satisfy $s \cdot s \cdot s \cdot s \cdot s = 1$ and $t \cdot t = 1$. The element $s \cdot t$ satisfies $s \cdot t \cdot s \cdot t \cdot s \cdot t = 1$. This is usually expressed by saying that s has order 5, t has order 2, and $s \cdot t$ has order 3. Show that the Cayley graph of A_5 with respect to the generators s and t looks like the pattern on a soccer ball illustrated in Figure D.5.

The order of an element g in a group G is the least positive integer n such that $g^n = 1$; the order is ∞ if there is no such n.

Fig. D.5 Cayley graph of A_5.

In Figure D.5, name one of the vertices 1. Then $s^5 = 1$ gives rise to a pentagon at that vertex, and $(st)^3 = 1$ gives rise to a hexagon at that vertex. Apply the definition of a Cayley graph to see how the other pentagons and hexagons arise. The pentagons and hexagons meet along edges because $t^2 = 1$.

Project D.27. Draw the Cayley graph for $\mathbf{Z}$ for the set of generators $\{1,2\}$.

Figure D.6 gives an indication of the Cayley graph of $\mathrm{PSL}_2(\mathbf{Z})$ with respect to the generators $[A]$ and $[B]$, where the matrices A and B are defined in Example D.18.

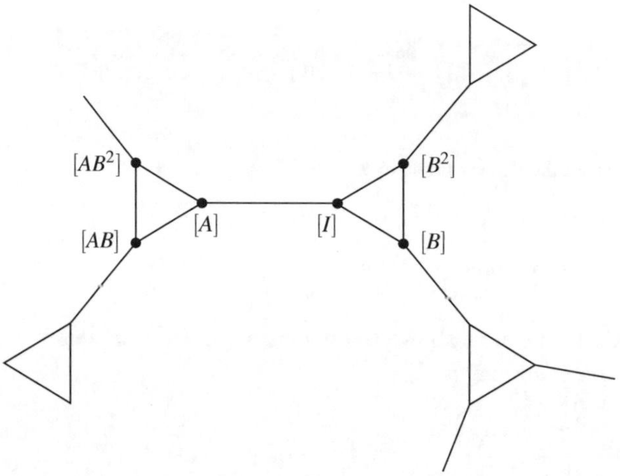

Fig. D.6 Part of the Cayley graph of $\mathrm{PSL}_2(\mathbf{Z})$.

And Figure D.7 gives an indication of the Cayley graph of $\mathrm{SL}_2(\mathbf{Z})$ with respect to the generators A and B.

D.7 G as a Group of Symmetries of Γ

The technical term is that h acts as an automorphism of the graph Γ.

Let G be a group finitely generated by S. Then G can be seen to be a group of symmetries of $\Gamma := \Gamma(G,S)$ as follows: Let $h \in G$. Then h "acts" on Γ by moving each vertex g to the vertex hg; since the vertices of Γ are the members of G, this makes sense. Note that h is multiplying on the left, whereas the rule for edges was given in terms of right multiplication by generators. A consequence of this is that whenever g_1 and g_2 span an edge, hg_1 and hg_2 also span an edge, so h takes edges to edges in a manner compatible with its action on vertices.

We close with some comments:

A finitely generated group has more than one Cayley graph: the Cayley graph depends on which finite generating set is chosen.

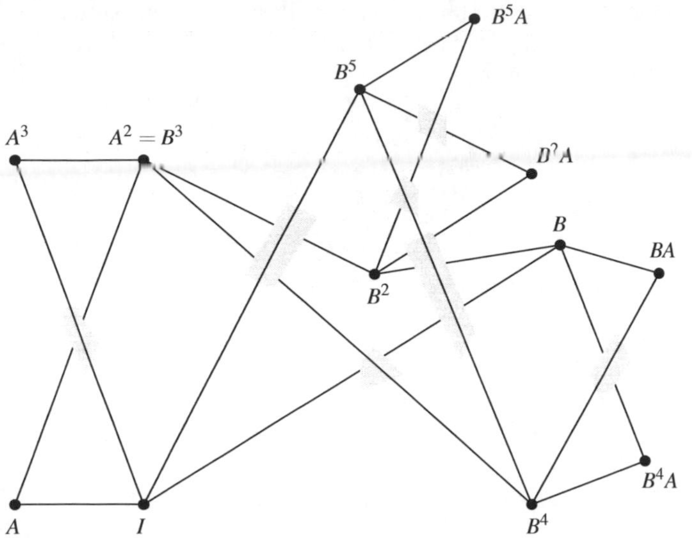

Fig. D.7 Part of the Cayley graph of $\mathrm{SL}_2(\mathbf{Z})$.

There are important invariants of a finitely generated group that can be computed from any Cayley graph, independent of which generating set is chosen; one of these is the "number of ends" of the group.

Usually, the group G is not the full group of symmetries of its Cayley graph, but just a subgroup that has the property that every vertex is "moved" by every member of G (except by 1).

This says that G acts freely *on its Cayley graph.*

D.8 Lie Groups

Everything in this section has concerned what are sometimes called "discrete groups." There are also "continous groups," the nicest of which are called *Lie groups*. Here are two examples:

Lie groups are named in honor of Sophus Lie (1842–1899).

Example D.28. The groups $\mathrm{SL}_2(\mathbf{R})$ and $\mathrm{PSL}_2(\mathbf{R})$ are defined in the same way as $\mathrm{SL}_2(\mathbf{Z})$ and $\mathrm{PSL}_2(\mathbf{Z})$, but using matrices with real number entries rather than just integer entries.

As sets, Lie groups are uncountable. But just as the techniques of analysis and topology take over from algebra when one deals with the whole set of real numbers (as distinct from a discrete subset like the integers), so analysis and topology mix with algebra in the study of Lie groups to give deep and elegant mathematics.

Among other applications, physicists describe elementary particles in terms of the symmetries exhibited by the laws defining those particles, and typically these groups of symmetries are Lie groups.

Appendix E
Generating Functions

The full beauty of the subject of generating functions emerges only from tuning in on both channels: the discrete and the continuous.
Herbert Wilf (*generatingfunctionology*, A K Peters, 2005, p. vii)

Assume $(a_n)_{n=0}^{\infty}$ is a sequence of real numbers that is not explicitly defined by a formula, for example, a recursive sequence. One can sometimes get useful information about this sequence—identities, formulas, etc.— by embedding it in a **generating function**:

$$A(x) := \sum_{n=0}^{\infty} a_n x^n .$$

We use the convention that the members of the sequence are named by a lowercase letter and the corresponding generating function is named by its uppercase equivalent.

We think of this series $A(x)$ as a *formal power series*, in the sense that questions of convergence are ignored. So operations on formal power series have to be defined from scratch.

E.1 Addition

When $A(x) = \sum_{n=0}^{\infty} a_n x^n$ and $B(x) = \sum_{n=0}^{\infty} b_n x^n$ are generating functions, we define their **sum** as

$$A(x) + B(x) := \sum_{n=0}^{\infty} (a_n + b_n) x^n ,$$

and multiplicaton by x as

$$x \sum_{n=0}^{\infty} a_n x^n := \sum_{n=0}^{\infty} a_n x^{n+1} .$$

Example E.1. Define a sequence $(a_n)_{n=0}^{\infty}$ recursively by

$$a_0 = 0 \qquad \text{and} \qquad a_{n+1} = 3a_n + 2 \qquad \text{for } n \geq 0 .$$

M. Beck and R. Geoghegan, *The Art of Proof: Basic Training for Deeper Mathematics*, Undergraduate Texts in Mathematics, DOI 10.1007/978-1-4419-7023-7_19,

Then $a_n = 3^n - 1$.

This is easily proved by induction on n. However, the proof given here shows how generating functions can be useful.

Proof. As before, define

$$A(x) = \sum_{n=0}^{\infty} a_n x^n.$$

Then the recurrence $a_{n+1} = 3a_n + 2$ gives rise to the generating-function identity

$$\sum_{n=0}^{\infty} a_{n+1} x^n = 3 \sum_{n=0}^{\infty} a_n x^n + 2 \sum_{n=0}^{\infty} x^n. \tag{E.1}$$

Proposition 12.2 says that this geometric series converges when $|x| < 1$, so for now, you can think of x as limited to that range. But in Section E.2 this equality will have a formal meaning without discussion of the admissible values of x.

The first summand on the right-hand side is $3A(x)$, and the second summand is a geometric series, which by Proposition 12.2 equals $\frac{2}{1-x}$. What about the term on the left-hand side? We can do a little algebra to conclude that

$$\sum_{n=0}^{\infty} a_{n+1} x^n = \frac{1}{x} \sum_{n=0}^{\infty} a_{n+1} x^{n+1} = \frac{1}{x} \sum_{n=1}^{\infty} a_n x^n.$$

However, $a_0 = 0$, so the expression on the right is just $\frac{1}{x}A(x)$. Thus (E.1) simplifies to

$$\frac{1}{x} A(x) = 3A(x) + \frac{2}{1-x},$$

from which it follows that

$$A(x) = \frac{2x}{(1-x)(1-3x)}.$$

The method of partial-fraction decomposition (known to all calculus students) gives

$$A(x) = -\frac{1}{1-x} + \frac{1}{1-3x}.$$

The two summands on the right can be converted back into (geometric) series:

$$A(x) = -\sum_{n=0}^{\infty} x^n + \sum_{n=0}^{\infty} (3x)^n = \sum_{n=0}^{\infty} (3^n - 1) x^n.$$

Remembering that $A(x) = \sum_{n=0}^{\infty} a_n x^n$, we conclude that $a_n = 3^n - 1$. □

Project E.2. Give a generating-function proof of Proposition 4.29. Compare your proof with that of Project 11.26. Generalize your proof along the lines of Proposition 11.25.

Project E.3. Compute the generating function $G(x) = \sum_{n=0}^{\infty} g_n x^n$ for the sequence $(g_n)_{n=0}^{\infty}$ defined recursively by

$$g_0 = g_{34} = 0 \qquad \text{and} \qquad g_{n+2} = g_{n+1} + g_n \qquad \text{for } n \geq 0\,.$$

E.2 Multiplication and Reciprocals

You have long known how to multiply two polynomials:

Proposition E.4. *The product of the two polynomials*

$$A(x) = a_d\,x^d + a_{d-1}\,x^{d-1} + \cdots + a_0$$

and

$$B(x) = b_d\,x^d + b_{d-1}\,x^{d-1} + \cdots + b_0$$

is

$$A(x)\,B(x) = c_{2d}\,x^{2d} + c_{2d-1}\,x^{2d-1} + \cdots + c_0\,,$$

where for $0 \leq n \leq 2d$,

$$c_n = a_0 b_n + a_1 b_{n-1} + \cdots + a_n b_0 = \sum_{k=0}^{n} a_k b_{n-k}\,.$$

We do not require both polynomials to be of degree d; i.e., some leading coefficients may be zero.

This proposition motivates the following definition. The **product** of two generating functions $A(x) = \sum_{n=0}^{\infty} a_n\,x^n$ and $B(x) = \sum_{n=0}^{\infty} b_n\,x^n$ is defined to be

$$A(x)\,B(x) := \sum_{n=0}^{\infty} \left(\sum_{k=0}^{n} a_k b_{n-k} \right) x^n. \tag{E.2}$$

Definition (E.2) allows us to define the **reciprocal** of the generating function $A(x) = \sum_{n=0}^{\infty} a_n\,x^n$ as the generating function $B(x) = \sum_{n=0}^{\infty} b_n\,x^n$ such that

$$A(x)\,B(x) = 1\,.$$

The right-hand side is the generating function 1.

For example, this allows us to view the geometric series as a formal power series: since

$$(1-x)\left(1 + x + x^2 + x^3 + \cdots\right) = 1\,,$$

we see that the reciprocal of the geometric series is $1 - x$, that is,

$$\sum_{k \geq 0} x^k = \frac{1}{1-x}\,.$$

Proposition E.5. *The generating function $A(x) = \sum_{n=0}^{\infty} a_n\,x^n$ has a reciprocal if and only if $a_0 \neq 0$.*

Hint: use (E.2) to recursively compute the coefficients of the reciprocal of $A(x)$.

Project E.6. Compute the sequence $(a_k)_{k=0}^{\infty}$ that gives rise to the generating function

$$A(x) = \sum_{k \geq 0} a_k x^k = \left(\frac{1}{1-x}\right)^2$$

by viewing $A(x)$ as the product of two geometric series.

Project E.7. Given a sequence $(a_n)_{n \geq 0}$ with generating function $A(x) = \sum_{n=0}^{\infty} a_n x^n$, let

$$B(x) := \sum_{n=0}^{\infty} b_n x^n = \frac{A(x)}{1-x}.$$

Find a formula for b_n.

Project E.8. This project is about the binomial coefficients $\binom{n}{k}$.

(i) First assume that n and k are integers such that $0 \leq k \leq n$. Assume you did not know anything about $\binom{n}{k}$, except the relations

We found these relations in Corollary 4.20.

$$\binom{n}{k} = \binom{n-1}{k-1} + \binom{n-1}{k} \quad \text{and} \quad \binom{n}{0} = 1.$$

Compute the generating functions

$$B_n(x) := \sum_{k \geq 0} \binom{n}{k} x^k$$

(one for each $n \geq 0$) and use them to find a formula for $\binom{n}{k}$.

(ii) Convince yourself that your computation will be unchanged if n is allowed to be any real (or even complex) number and k to be any nonnegative integer.

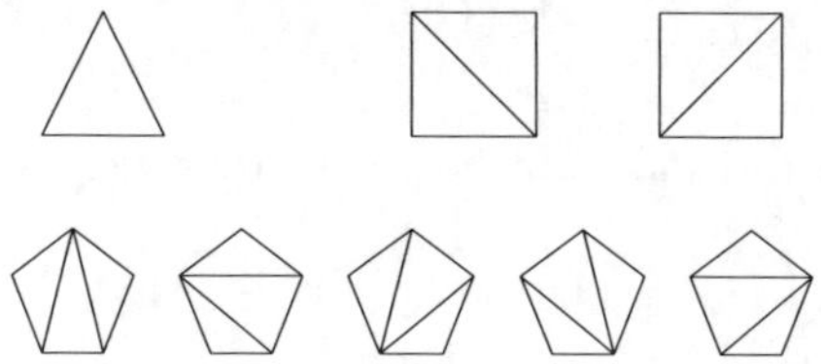

Fig. E.1 Illustration of the first three Catalan numbers.

The c_n are called Catalan numbers. *They occur in many different places in mathematics.*

Project E.9. Let c_n denote the number of triangulations of an $(n+2)$-gon, so for example, $c_1 = 1$, $c_2 = 2$, $c_3 = 5$, as illustrated in Figure E.1. We also set $c_0 = 1$.

(i) Find a recurrence relation for c_n.

(ii) Compute the generating function $C(x) := \sum_{n=0}^{\infty} c_n x^n$.

(iii) Use $C(x)$ and Project E.8 to derive a formula for c_n.

Project E.10. Use Projects E.2 and E.7 to prove that the Fibonacci numbers f_n satisfy

$$f_0 + f_1 + \cdots + f_n = f_{n+2} - 1 .$$

E.3 Differentiation

In the same spirit as before, we define the **derivative** of the generating function $A(x) = \sum_{n=0}^{\infty} a_n x^n$ to be

$$A'(x) := \sum_{n=1}^{\infty} n\, a_n x^{n-1} .$$

It is often useful to multiply a derivative by x:

$$x A'(x) = \sum_{n=1}^{\infty} n\, a_n x^n .$$

Project E.11. Define the sequence $(a_n)_{n=0}^{\infty}$ recursively by

$$a_0 = 0 \qquad \text{and} \qquad a_{n+1} = 3a_n + 2n \qquad \text{for } n \ge 0 .$$

Find a formula for a_n.

Project E.12. Find the sum of the first n squares by differentiating the geometric series (viewed as a generating function). Generalize.

Revisit Project 4.14 and our hint to that project.

Appendix F
Cardinal Number and Ordinal Number

Mathematics as we know it and as it has come to shape modern science could never have come into being without some disregard for the dangers of the infinite.
David Bressoud (*A Radical Approach to Real Analysis*, Mathematical Association of America, 2007, p. 22)

Recall from Chapter 13 that we write $\operatorname{card} A = \operatorname{card} B$ if there is a bijection $A \to B$. Then we say that A and B have the same cardinal number. We write $\operatorname{card} A \leq \operatorname{card} B$ if there is an injection $A \to B$. In this latter case we say that the cardinal number of A is less than or equal to the cardinal number of B. Recall that $\operatorname{card} \mathbf{N} \leq \operatorname{card} \mathbf{R}$ but $\operatorname{card} \mathbf{N} \neq \operatorname{card} \mathbf{R}$.

The intuitive idea is that cardinal number is a measure of the size of a set, i.e., a measure that generalizes to infinite sets the notion of "number of members" in the case of finite sets. But this raises two natural questions:

- If $\operatorname{card} A \leq \operatorname{card} B$ and $\operatorname{card} B \leq \operatorname{card} A$, is it true that $\operatorname{card} A = \operatorname{card} B$?
- Given two sets A and B, is at least one of the statements $\operatorname{card} A \leq \operatorname{card} B$ and $\operatorname{card} B \leq \operatorname{card} A$ true?

Without positive answers to both questions, cardinal number would not appear to be a good generalization of the notion of "number of members." Both questions do have positive answers, and our purpose in this chapter is to explain why.

F.1 The Cantor–Schröder–Bernstein Theorem

Theorem F.1. *If* $\operatorname{card} A \leq \operatorname{card} B$ *and* $\operatorname{card} B \leq \operatorname{card} A$*, then* $\operatorname{card} A = \operatorname{card} B$.

Proof. Let $f : A \to B$ and $g : B \to A$ be injections. Our goal is to create a bijection $A \to B$.

M. Beck and R. Geoghegan, *The Art of Proof: Basic Training for Deeper Mathematics*, Undergraduate Texts in Mathematics, DOI 10.1007/978-1-4419-7023-7_20,

The trick is to assign a "score"—a nonnegative integer or ∞—to each member of A and to each member of B. We do this in detail for B; the procedure for A is similar.

So consider some $b_0 \in B$. Recall that the preimage

$$f^{-1}(b_0) := \{a \in A : f(a) = b_0\}$$

consists of all those members of A that are mapped to b_0 by f. Because f is an injection, this set is either empty or it consists of exactly one point.

If $f^{-1}(b_0)$ is empty, b_0 gets score 0.

If $f^{-1}(b_0)$ contains one element, call it a_1, then we consider the set

$$g^{-1}(a_1) := \{b \in B : g(b) = a_1\}$$

consisting of those members of B that are mapped to a_1 by g. Because g is an injection, this set is either empty or it consists of exactly one point.

If $g^{-1}(a_1)$ is empty b_0 gets score 1.

If $g^{-1}(a_1)$ contains one element, call it b_2, then we consider the set

$$f^{-1}(b_2) := \{a \in A : f(a) = b_2\}$$

and proceed as before. Since there is clearly an inductive definition happening here, and we want to describe it informally, we will bore you by doing one more step:

The set $f^{-1}(b_2)$ consists of all those members of A that are mapped to b_2 by f. Because f is an injection, this set is either empty or it consists of exactly one point.

If $f^{-1}(b_2)$ is empty b_0 gets score 2.

If $f^{-1}(b_2)$ contains one element, call it a_3, then we consider the set

$$g^{-1}(a_3) := \{b \in B : g(b) = a_3\}$$

consisting of those members of B that are mapped to a_3 by g. And so on.

This discussion of assigning scores to members of B illustrates the good and bad sides of writing mathematics informally. The good side is that it is easier to see what is intended by reading the first few cases than by reading a base case, an induction hypothesis, and the next inductive definition. For the point b_0 we are defining, inductively, a sequence $b_0, a_1, b_2, a_3, b_4, \ldots$ and when it terminates the last subscript appearing is the score of b_0.

The bad side is that this informality may disguise the possibility that the sequence might not terminate. In that case b_0 is assigned the score ∞.

The element $a_0 \in A$ is given a score in a similar way: one produces a sequence $a_0, b_1, a_2, b_3, a_4, \ldots$, and the score for a_0 is either the last subscript occurring, or is ∞.

We now partition B into three sets B_E, B_O, and B_I: The set B_E consists of all points with even scores, B_O those with odd scores, and B_I those whose score is ∞. We do the same with A so that $A = A_E \cup A_O \cup A_I$, where these three subsets are pairwise disjoint.

We can now define a bijection $h : A \to B$.

(1) If $a_0 \in A_O \cup A_I$, then $g^{-1}(a_0)$ contains exactly one point b_1; in this case, define $h(a_0) = b_1$.

(2) If $a_0 \in A_E$, define $h(a_0) = f(a_0)$.

First, note that h is well defined; i.e., we have unambiguously stated what $h(a)$ is for every $a \in A$. In other words, h is a function.

Next, note that h maps A_O into B_E, A_I into B_I, and A_E into B_O. It follows that if a' and a'' lie in different sets among the trio of sets A_E, A_O, A_I then it cannot happen that $h(a') = h(a'')$. Moreover, it is clear from the definition that h is (separately) injective on each set of this trio: that is because $h = g^{-1}$ gives a well-defined injective function on $A_O \cup A_I$, and h agrees with the injective function f on A_E. It follows that h is injective.

To see that h is surjective, consider an arbitrary element $b_0 \in B$. If $b_0 \in B_O \cup B_I$ then b_0 does not have score 0, so $a_1 = f^{-1}(b_0)$ exists, and $h(a_1) = b_0$. And if $b_0 \in B_E$ then $a_1 = g(b_0) \in A_O$, so $h(a_1) = g^{-1}(g(b_0)) = b_0$. Thus, h is surjective. □

F.2 Ordinal Numbers

The second question posed at the beginning of this chapter was this: is at least one of the statements card $A \leq$ card B and card $B \leq$ card A true? In other words, are any two sets "comparable"? This requires a discussion of well-ordered sets.

A **well-ordering** on a nonempty set A is a relation $\leq$ satisfying

(i) $a \leq b$ and $b \leq a$ imply $a = b$;

(ii) $a \leq b$ and $b \leq c$ imply $a \leq c$;

(iii) $a \leq b$ or $b \leq a$ holds for any $a, b \in A$;

(iv) every nonempty subset of A has a least member with respect to $\leq$.

For example, by Theorem 2.32, the usual less-than-or-equal ordering is a well-ordering of $\mathbf{N}$.

The $\leq$ ordering on $\mathbf{R}$ is not a well-ordering on $\mathbf{R}$. Why?

A **well-ordered set** is a set equipped with a well-ordering. When $(A, \leq)$ is a well-ordered set it is customary to omit the $\leq$ and to say "A is a well-ordered set." We denote the least element of A by 1_A. The set of all members of A that are $\leq a \in A$ is denoted by $[1_A, a]$. Similarly, $[1_A, a)$ denotes the set of all elements of A that are strictly below a in terms of the given order.

Imitating the interval notation you are familiar with for numbers, we call $[1_A, a]$ a closed interval *in* A *and* $[1_A, a)$ a half-open interval *in* A.

Every nonempty subset $C \subseteq A$ inherits a well-ordering from A; i.e., just use the same $\leq$ on C.

Now let A and B be nonempty well-ordered sets. For simplicity of notation we denote both well-orderings by $\leq$; context will indicate which ordering is meant at any particular moment. We will *try* to construct an order-preserving injection $A \to B$, i.e., an injective function $f : A \to B$ such that

$$a \leq a' \qquad \text{implies} \qquad f(a) \leq f(a').$$

There may not be such an injective function—just think of the case $A = \{1,2,3\}$ and $B = \{1,2\}$—but in this case there is an injective function in the other direction. As we will see, that is what happens in general.

Attempting to define an order-preserving injection $A \to B$, we begin by defining $f(1_A) := 1_B$.

Now, imitating the idea of induction, assume that $f(a)$ has been defined for all $a \in [1_A, a_0)$. Next (and here is the dangerous moment) we define $f(a_0)$ to be the least element of the subset of B not already in the image of f; i.e., we define $f(a_0)$ to be the least element of $B - f([1_A, a_0))$. The danger is that $B - f([1_A, a_0))$ might be empty, in which case it would not have a least element. There are two cases:

(i) for every $a_0 \in A$, $B - f([1_A, a_0))$ is nonempty;

(ii) for some $a_0 \in A$, $B - f([1_A, a_0))$ is empty; i.e., $f([1_A, a_0)) = B$.

In the first case, we have constructed the desired injection $f : A \to B$, provided our process of imitating induction is legitimate mathematics. Recall that the legitimacy of induction in the well-ordered set **N** follows from Axiom 2.15. But the set A might be uncountable, so we cannot rely on that axiom here. We must tell you that in standard set theory our process of induction on well-ordered sets is considered legitimate: whether that sentence is an axiom or is derived from more primitive axioms depends on how set theory is being presented; these matters are outside the scope of this book. This generalized form of induction is called *transfinite induction*.

In the second case, the function f maps $[1_A, a_0)$ injectively *onto* B. Hence it defines a bijection. Its inverse gives an injection from B into A. Thus we have proved the following theorem.

Theorem F.2. *Given two (nonempty) well-ordered sets there is an order-preserving injection of one into the other.*

If A and B are well-ordered sets we say that ord $A \leq$ ord B (the ordinal number of A is less than or equal to the ordinal number of B) if there is an order-preserving injection $A \longrightarrow B$. Well-ordered sets A and B **have the same ordinal number** if there is an order-preserving bijection $A \longrightarrow B$.

Now, an axiom of set theory says that

every nonempty set admits a well-ordering.

By this axiom, **R** *admits a well-ordering. However, it is impossible to explicitly describe a well-ordering on* **R**.

This is a strong statement to admit into mathematics, and there is much to be said about it. All we will say here is that it is logically equivalent to a generally accepted axiom of set theory called the Axiom of Choice.

We have mentioned the Axiom of Choice once before: in the proof of Proposition 9.10.

From this axiom of set theory we conclude the following theorem.

Theorem F.3. *Given two sets A and B, at least one of the statements*

$$\operatorname{card} A \leq \operatorname{card} B \qquad \textit{and} \qquad \operatorname{card} B \leq \operatorname{card} A$$

is true.

Proof. If one of the sets is empty, this is trivial. Otherwise, put a well-ordering on each and apply Theorem F.2. □

Appendix G
Remarks on Euclidean Geometry

The purely formal language of geometry describes adequately the reality of space. We might say, in this sense, that geometry is successful magic. I should like to state a converse: is not all magic, to the extent that it is successful, geometry?
Rene Thom (*Structural Stability and Morphogenesis*, W. A. Benjamin, Reading, MA, 1975, p. 11)

You have studied geometry in high school, a version of what was assembled from knowledge of the ancient Greeks (around 300 BCE) by the Greek-speaking textbook writer Euclid, who lived in Alexandria, Egypt.

You may have been taught geometry from a practical-life point of view, where the most important thing was to have an intuitive grasp of what is true about lines, angles, circles, etc. But Euclid understood what nowadays is called the Axiomatic Method, and he attempted to present geometry much in the manner we used in this book to present the integers and the real numbers.

There were the undefined terms *point, straight line, finite straight line, angle, right angle, circle, congruent.* The axioms stated by Euclid can be given in modern language as follows:

1. If A and B are distinct points there is a unique finite straight line joining A to B.
2. A finite straight line can be extended to be a straight line, and this straight line is unique.

Euclid never asserted uniqueness explicitly, but took it for granted in his work.

3. Given a point A and a finite straight line l, there is a unique circle having A as center and l as radius.
4. All right angles are congruent.
5. If two lines l_1 and l_2 are drawn that intersect a line l_3 in such a way that the sum of the inner angles on one side is less than two right angles, then l_1 and l_2 intersect each other on that side if extended far enough.

This fifth axiom is called the parallel postulate, *and there is much to be said about its place in geometry.*

We make no attempt here to develop this subject. There are many good sources to be found in books and on the Internet. We simply point out that if geometry is to be done according to the standards of modern mathematics, all theorems of geometry

M. Beck and R. Geoghegan, *The Art of Proof: Basic Training for Deeper Mathematics*, Undergraduate Texts in Mathematics, DOI 10.1007/978-1-4419-7023-7_21,

must be deduced from these axioms or from previous theorems that were deduced from these axioms.

List of Symbols

The following table contains a list of symbols that are frequently used throughout this book. The page numbers refer to the first appearance/definition of each symbol. Those symbols having definitions both in **Z** and in **R** come with two page numbers.

Symbol	Meaning	Page
$=$	equals	5
$=$	equals (as sets)	17
$\neq$	is not equal to	5
$:=$	is defined by	18
$+$	addition	4, 76
$-$	subtraction	10, 78
$-$	set difference	51
$\cdot$	multiplication	4, 76
$/$	division	78
$\mid$	divides	7
$\in$	is an element of	5
$\notin$	is not an element of	5
$\subseteq$	is a subset of	14
$\cap$	intersection (of sets)	50
$\cup$	union (of sets)	50
$\times$	Cartesian product (of sets)	52
$<$	is less than	15, 79
$>$	is greater than	15, 79
$\leq$	is less than or equal to	15, 79
$\geq$	is greater than or equal to	15, 79
$\equiv$	is congruent to	60
$\sim$	is related to	56
$-m$	additive inverse of m	4, 77
x^{-1}	multiplicative inverse of x	77
m^2	$m \cdot m$	17
m^3	$m^2 \cdot m$	19
m^k	m to the power k	36

Symbol	Meaning	Page
0	zero	4, 76
1	one	4, 76
2	two	7
3	three	19
$\sqrt{2}$	square root of 2	103
$\sqrt{r}$	square root of r	104
$\sqrt[n]{r}$	n^{th} root of r	110
π	ratio of the circumference of a circle to its diameter	128
i	the complex number $(0,1)$	146
$\|x\|$	absolute value of x	57, 97
$\|x-y\|$	distance from x to y	98
$(x_k)_{k=1}^{\infty}$	a sequence $x_1, x_2, x_3 \ldots$	34
$\lim_{k\to\infty} x_k$	limit of of the sequence $(x_k)_{k=1}^{\infty}$	99
$\sum_{j=1}^{k} x_j$	a (finite) sum	35
$\sum_{j=m}^{n} x_j$	another sum	37
$\prod_{j=1}^{k} x_j$	a product	35
$k!$	k factorial	35
$\binom{k}{m}$	binomial coefficient "k choose m"	39
$\sum_{j=1}^{\infty} a_j$	an infinite series	114
min	smallest element (of a set of numbers)	21, 83
max	largest element (of a set of numbers)	82
sup	smallest upper bound (of a set of numbers)	82
inf	greatest lower bound (of a set of numbers)	83
gcd	greatest common divisor (of two integers)	22
$\nu(n)$	number of digits of n	66
$\mathrm{Re}(z)$	real part of z	146
$\mathrm{Im}(z)$	imaginary part of z	146
$e^{i\phi}$	$\cos\phi + i\sin\phi$	148
$m.d_1d_2d_3\ldots$	decimal representation	116
$f: A \to B$	a function with domain A and codomain B	53
id_A	identity function on A	86
$g \circ f$	composition of f and g (first f then g)	87
$f^{-1}(C)$	preimage of C	136
$\varnothing$	empty set	50
$P(A)$	power set of the set A	126
S_n	set of all permutations of $[n]$	152
$\Gamma(G,S)$	Cayley graph of G generated by S	156
$\forall$	for all	26
$\exists$	there exist(s)	26
$\exists!$	there exist(s) unique	27
$\Rightarrow$	implies	28
$\Leftrightarrow$	if and only if	28
$\square$	end of a proof	5
$[a]$	equivalence class of a	56
$[x,y]$	(closed) interval from x to y	82

Symbol	Meaning	Page
$[n]$	the set $\{1,2,\ldots,n\}$	122
$\mathbf{N}$	the set of natural numbers	14
$\mathbf{Z}$	the set of integers	4
$\mathbf{Z}_{\geq 0}$	the set of nonnegative integers	35
$\mathbf{Q}$	the set of rational numbers	108
$\mathbf{R}$	the set of real numbers	76
$\mathbf{R}_{>0}$	the set of positive real numbers	79
$\mathbf{C}$	the set of complex numbers	145

Index